Lean Six Sigma
Statistics

Other Books in the Six Sigma Operational Methods Series

Lean Six Sigma
Statistics

Calculating Process Efficiencies
in Transactional Projects

Alastair K. Muir, Ph.D.

President
Muir & Associates Consulting, Inc.
Calgary, Alberta, Canada

McGraw-Hill

New York   Chicago   San Francisco   Lisbon   London   Madrid
Mexico City   Milan   New Delhi   San Juan   Seoul
Singapore   Sydney   Toronto

The **McGraw·Hill** Companies

Cataloging-in-Publication Data is on file with the Library of Congress

Copyright © 2006 by The McGraw-Hill Companies, Inc. All rights reserved. Printed in the United States of America. Except as permitted under the United States Copyright Act of 1976, no part of this publication may be reproduced or distributed in any form or by any means, or stored in a data base or retrieval system, without the prior written permission of the publisher.

1 2 3 4 5 6 7 8 9 0   DOC/DOC   0 1 0 9 8 7 6 5

ISBN 0-07-144585-4

The sponsoring editor for this book was Kenneth P. McCombs and the production supervisor was Richard C. Ruzycka. It was set in Times by International Typesetting and Composition. The art director for the cover was Handel Low.

Printed and bound by RR Donnelley.

 This book is printed on recycled, acid-free paper containing a minimum of 50% recycled, de-inked fiber.

McGraw-Hill books are available at special quantity discounts to use as premiums and sales promotions, or for use in corporate training programs. For more information, please write to the Director of Special Sales, McGraw-Hill Professional, Two Penn Plaza, New York, NY 10121-2298. Or contact your local bookstore.

CONTENTS

PREFACE

Six Sigma is a proven, successful methodology for solving business problems with customer focus and quantifiable financial benefits to the business. Early adopters expanded and applied Six Sigma to areas outside manufacturing, where it first began. As we continued to apply Six Sigma to transactional processes such as order management, new business acquisition and integration, accounts payable and accounts receivable functions, we ran into the same technical problems over and over again. We have modified the toolkit and the logic beneath Six Sigma to focus on projects where the emphasis is on understanding cycle time in a transactional environment.

The Development of This Book

It is interesting to look back on how this book evolved from an idea to the printed volume you now hold. Different sections of this book have been written over the years in response to common problems faced by *Black Belts* (BBs) and *Master Black Belts* (MBBs) working in a wide variety of industries. Each individual issue would solicit a single response, sometimes quick and concise, often more extended and building on basic Six Sigma concepts.

Over the years, these issues continued to occur and began to cluster into groups with their own characteristics and solutions. We assembled some of the more extended topics into advanced MBB training courses at the GE corporate training facility in Crotonville, New York, and elsewhere around the globe. Some of the less involved responses were presented at conferences and in a series of articles written for the Six Sigma website, www.iSixSigma.com.

When McGraw-Hill first approached us about writing this book, we tried to compile these incremental changes into a single book as an augmentation of Six Sigma. We soon learned that the existing Six Sigma body of knowledge required a substantial redesign to address the types of problems in transactional business processes. The core of the book became one about the logic and tool usage for executing transactional Six Sigma projects broken into the five phases of a *define, measure, analyze, improve, and control* (DMAIC) project.

As the book reached the first review phase, we scored it against the *quality function deployment* (QFD) constructed during the design sessions for the book. It showed that it was still not complete in addressing some of the issues about project selection and managing a portfolio of projects directed toward a bigger problem. For those who have done this, we found we had blank rows in the QFD. When this occurred, it meant we had no deliverables that addressed issues important to the reader audience. These problems clustered into areas related to the implementation of a Six Sigma program rather than the execution of individual Six Sigma projects.

We decided to draw on our years of experience with successful and unsuccessful aspects of Six Sigma implementations at various business units of GE, Bombardier, Air Canada, Bank of America, and government agencies. One of the biggest differences between the two groups of implementations was the level of direction given to the BBs and MBBs from the executive level of the company. A lack of data driven rigor for management buy-in would lead to a lukewarm program with few long-term benefits and little customer impact.

At Muir & Associates Consulting, we augmented our Six Sigma program to address these systemic problems. The data driven nature parallels the philosophy of Six Sigma, but lies outside of the usual DMAIC project regime. We included enough of this material in the book to increase the success of those BBs and MBBs reading it. The Recognize and Sustain phases will not be a part of your DMAIC projects, but should be present in your Six Sigma program. You can think of this as the bigger picture of your company's Six Sigma program where individual DMAIC projects become the Improve phase of the Six Sigma program. An important part of a successful implementation for you is a clear definition of the communication plan and the roles and responsibilities of the BBs, MBBs, quality leaders, and executive. While we did not want this book to become a book about implementation, we wanted to include enough of the scaffold of the infrastructure to make you more effective in understanding the logic behind project selection and reporting.

Background Required

The senior editor and I spent a long time discussing the needs of the practicing BBs or MBBs who have come from a manufacturing background, and are starting to tackle more complex problems using the familiar DMAIC methodology. It is assumed that you have taken Six Sigma training and are familiar with the DMAIC methodology in a broad sense. We do not spend a lot of ink discussing existing tools, such as t-tests or SWOT analysis, except where we use them in an unfamiliar manner. It is not our intention to summarize all the details of topics such as hypothesis testing, the statistical details of alpha versus beta risk, or quantitative risk assessment, but to present the idea at the opportune moment.

We assume you know more about Six Sigma than you do about lean manufacturing. A subset of concepts from lean manufacturing has been incorporated into the Six Sigma DMAIC project regime where it is appropriate.

When the results of projects are being presented to the stakeholders and project sponsors, there is sometimes an overemphasis on tool usage. We have learned over the years that emphasizing technical brilliance while neglecting the project management aspects can lead to disaster or project failure. Six Sigma is all about answering questions. We have developed the flow in this book based on answering questions first, and showing the details of the tools second.

Summary of Chapter Contents

No one can read a book from cover to cover and remember all the details at exactly the right time during a project. We have broken up the topics and introduced them in the context of the different phases of a DMAIC project. Each chapter ends with a checklist of questions that are appropriate at that stage. Resist the temptation to skip sections during your project. For example, in transactional projects, we often see project teams accept the vast amount of data that is available to them and charge ahead to the Analyze phase. Keep in mind that the audience will be interested in the source and reliability of your data in a broad sense, and will ignore your analysis if they are not satisfied that the data is reliable and representative of the business process.

Chapters 1 and 2

Transactional processes are different from manufacturing processes in many aspects. The data is different, there is little physical inventory and each transaction can be substantially different from the previous one. The Six Sigma program has evolved over the years to include the tools from other disciplines such as "lean manufacturing" to execute transactional projects. There is no best way for businesses to be. Businesses can be designed to be low volume, high margin, flexible custom shops or high volume, low margin, rigid assembly lines. The market and your customers will determine which is the best balance for your business. Your business will probably have elements of each for different service and customer segments.

Chapters 3 and 9

These two chapters, Recognize and Sustain, define what you can expect of the executive and the Six Sigma program custodian. The Six Sigma program custodian will be the person who is responsible for setting the direction of the program and reporting on the health and success of the entire initiative. They should not be involved in projects except at the identification and sponsorship level. At the DMAIC project level, you should be aware of how a decrease in cycle time will impact the financial statements of the company. The program custodian is responsible for designing a robust communication plan that will allow BBs to operationalize elements of the strategic plan and allow the executive to report the success of projects against those goals.

Chapter 4—Define

This is the first chapter you should be following as you begin your project. The real goal of a successful company is to satisfy its customers. This requires two elements to come together, the customer's expectation for the cycle time of a transaction, and your capacity to match that expectation. The primary metric is *variance to customer want* (VTW), measured by span. Role and responsibilities are assigned during the project scoping session. The resistance plan is combined with the stakeholder communication plan.

Chapter 5—Measure

When you are doing a project, you will always spend much more time in the Measure phase than you plan. Data is never what you expect, and it is never as reliable as purported by its owners. Consolidated data can be used to hide detail that is important to the customer, but may make some people look bad. We have replaced the traditional Gage R&R with the more general "data audit." You will get some amazing results after conducting a good review of data sources and operational definitions of what will seem like obvious terms. Many problems in business occur simply because no one is aware of them using the existing data and reports.

Chapter 6—Analyze

The general definition of a defect in lean Six Sigma is, "whenever there is a difference between the expected and actual time to complete a transaction without error." This means that your analysis will be centered on time measurements. The distributions of the data will never be normally distributed. These deviations from normality make it difficult or impossible to apply the statistical tools covered in basic Six Sigma training. We demonstrate the best tools to examine sets of nonnormal distributions in order to identify the key factors causing customer dissatisfaction. We look at the effect of handling different requests with different priorities as seen in medical triage and preferred customer programs.

Chapter 7—Improve

Once you have determined the sources of variation, then there are some solutions and improvement techniques based on lean manufacturing that are easy to apply. Your solution will not be a purely lean one; there is an optimum balance between "make to order" and "make to inventory." Process simulation is used to test the consequences of making changes to a dynamic and nonlinear process that may be separated in geographical and temporal terms. Kanban size calculations should be done in the presence of variation in delivery time, risk, and order quantity using Monte Carlo simulations.

Chapter 8—Control

There are three philosophies for the Control phase: mistake proofing, monitoring, and risk management. We cover some common and effective techniques of mistake proofing for transactional projects. We also introduce quantitative methods for risk assessment with uncertainty in a similar manner to the Kanban size calculation in Chap. 7.

Appendix A—Quantitative Risk Assessment

Crystal Ball is an Excel add-on that incorporates Monte Carlo simulation into spreadsheets. We demonstrate the construction of two models to identify the key steps effecting cycle time for an insurance underwriting process and a medical claims process. The data are drawn from two successful Six Sigma projects.

Appendix B—Process Simulation

It is difficult to predict the effects of changes in nonlinear systems and processes. Setting up a process for simulation is much more detailed than the process mapping done during the Define or Measure phase of a project. We show the general flow for constructing a model using the process simulation software—ProcessModel. Specific variables must be declared and manipulated to yield the information we require for testing improvement strategies.

Appendix C—Statistical Analysis

In lean Six Sigma, we use some statistical tools not available in the usual office software applications. While there are many statistical software packages available, we have selected Minitab as the best compromise between functionality, standardization, and user friendliness. We show the basic interface and introduce the extensive, built in help system of the package.

Alastair K. Muir, Ph.D.

ACKNOWLEDGMENTS

There can be no solutions without problems. If this book is a success at helping you with your project, it is only because business leaders have presented me with a variety of fascinating problems. A number of people have been particularly helpful by challenging and encouraging me:

Michael R. Chilton, President and CEO, Xin Hua Control Engineering, GE Energy, Shanghai, China.

Rose Marie Gage, Senior Manager, Corporate Initiatives, GE, Mississauga, Ontario, Canada.

Piet van Abeelen, Corporate Vice President for Six Sigma, GE, Fairfield, Connecticut, U.S.A.

Frank B. Serafini, Senior Vice President and Quality Executive, Bank of America, Charlotte, North Carolina, U.S.A.

Joe Carrabba, President, Diavik Diamond Mines Inc., Yellowknife, NWT, Canada.

Tom Rohleder, Professor, Haskayne School of Business, University of Calgary, Calgary, Alberta, Canada.

James de Vries, Senior Consultant and MBB, Six Sigma Academy, Houston, Texas, U.S.A.

Greg Stevenson, Investment Executive, ScotiaMcLeod, Calgary, Alberta, Canada.

Ken McCombs, Senior Acquisitions Editor, McGraw-Hill Professional, New York, New York, U.S.A.

Lean Six Sigma
Statistics

Manufacturing and Transactional Processes

1.1 Moving from Manufacturing to Services

Your CEO has just read a book on the benefits of the application of *lean manufacturing* in a service based business. Now convinced of the need for this in your business, it has fallen to you, the VP - operations, quality leader, *Master Black Belt* (MBB), *Black Belt* (BB), or other quality professional, to execute a few projects and generate benefits to the customers with a financial impact on the business. This book is designed to help you determine what you are going to do tomorrow morning.

The improvement projects you have been a part of until now have been focused on manufacturing processes. You have access to a large number of completed projects and have seen the application of a variety of Six Sigma tools.

Manufacturing processes are fairly good already, so an improvement project starting with a process that is already at 95 to 99 percent good is typical. The defect reduction is in the form of reduced scrap and rework costs or warranty returns.

Processes that are transactional, such as order management, quote and order preparation, credit checking, mortgage application processing, project scheduling, medical testing, and communication of engineering change notices are dependent on human communication, and are typically very poor to begin with. It is not unusual to find that 30 to 50 percent of the transactions require some kind of rework. The good news is that transactional processes can usually be greatly improved with very little capital outlay.

1.2 This Book

The tools and project management milestones presented here are based on the author's years of experience in executing and mentoring thousands of Six Sigma projects at GE and other *Fortune* 100 companies. This book will help you avoid the traps and dead ends you will undoubtedly meet.

The early design for this book was sketched out using a Define tool, In Frame–Out of Frame (Fig. 1.1). Our emphasis is on the practical aspects of executing a lean Six Sigma project based on the *Define, Measure, Analyze, Improve, and Control* (DMAIC) project discipline. The portions of the project familiar to you, such as surveying the *voice of the customer* (VOC),

Out of frame

In frame

Lean Six Sigma DMAIC project execution	Applications in transactional environments	Basic elements of the annual report Non-normal data
Realistic goals for process entitlement	Triage and queue management	Prioritization of business problems for project selection
Financial risk assessement and analysis	Failures and lessons learned Resistance	Probability plots Logistic regression
Transactional Gage R&R and the data audit	Roles and responsibilities in lean Six Sigma projects	Hazard plots to determine the time dependence of failure

Review of Lean Six Sigma success stories from the business press	Business philosophy Business transformation	Lean Six Sigma implementation Endless testimonials from fortune 500 CEOs	Advertisement for consulting services

Figure 1.1 In Frame–Out of Frame for the Book Design

conducting a stakeholder analysis, and designing a project communication plan, will be briefly reviewed where they are relevant. More detail will be presented in the newer areas, such as non-normal data analysis and financial risk assessment.

Cycle time reduction projects require data analysis you have probably not encountered in your previous DMAIC projects. The methods and tools presented in this book have been developed specifically to deal with the unique qualities of your transactional project.

1.3 The Effect of Variation on the Customers and Shareholders

Consider a production facility where daily production is shipped overnight to the customer (Fig. 1. 2). A machine breakdown on Monday resulted in an 80 percent reduction of the daily production. For the next few days the management ran the factory at 120 percent capacity to make up for the loss. If the overtime premium is 50 percent, then the production cost will increase by 24 percent for the week, even though the weekly total and

	Consistent production		Inconsistent production	
	Producer's production (units)	Customer's production cost ($)	Producer's production cost	Customer's production cost ($)
Monday	100		20	
Tuesday	100	10,000	120	10,000
Wednesday	100	10,000	120	$10,000 + 1.5 \times 2,000$
Thursday	100	10,000	120	$10,000 + 1.5 \times 2,000$
Friday	100	10,000	120	$10,000 + 1.5 \times 2,000$
Saturday		10,000		$10,000 + 1.5 \times 2,000$
Total	500	50,000	500 units	62,000
Average	100/day	10,000/day	100 units/day	12,400/day
Span	0/day	0/day	80 units/day	2,400/day

Figure 1.2 Effect of Variation in Supply on the Downstream Customer

daily average for production look fine. The machinery and people have also been worked overtime, thereby increasing the probability of another breakdown.

The problems caused by Monday's breakdown are not restricted to the producer. When the delivery on Tuesday morning was short by 80 percent, the customer's factory was filled with workers expecting to work a full shift. On the following four days when the producer delivered a shipment larger than the usual 100 units, the customer was required to pay the workers overtime to clear out the raw material. The customer may also not be able to physically or financially accept a larger than usual delivery. The sourcing department at the customer's site may have negotiated a price reduction on the extra units or applied liquidated damages to their loss of production on Tuesday.

In this example, the production metrics are reported at the corporate level at the end of each week. This consolidation has hidden Monday's production problem from senior management. The customers have complained to the senior management about supply problems, while it appears that the production facility met its weekly production target. The finance department was probably the last group to find out that the invoices and payments were irregular for the week.

Business culture can run contrary to the drive for consistent and predictable output. When a VP of production's compensation is based on achieving or exceeding production targets, it can create a business culture hostile to focusing on the customer. Managers can also feel more "in-touch" with their customers if they spend a large amount of their time expediting orders, and identifying production and delivery problems on an incident by incident basis. The illusion is that they are focused on the customer by overemphasizing short-term tactics over long-term planning. The best run businesses can best be described as boring. These businesses are run by managers who understand changes in customer demand and their own production capacity.

In a transactional environment, variation in cycle time can result in dissatisfied customers or loss of business. For a moment consider that you are thinking of buying a house and are shopping for a mortgage. You have approached two different banks and the rates, terms and setup costs are identical for both offerings. The local branches of the two banks will negotiate the deal, gather credit and legal information, send the application to the regional underwriting center where the mortgage will be approved or not, then make arrangement for the transfer of funds to the local branch.

There can be a number of reasons why this may proceed smoothly or not, but these do not concern you now.

You ask one more question of both banks, "How long does it take to have access to the funds?"

The two banks respond:

- Clare's House O' Money can complete the process in an average of 10 working days.
- Francis' Bank-o-rama can complete the process in an average of 8 days.

You will probably choose the latter. Introduce the idea of measuring variation and ask the same question. The responses now are:

- Clare's House O' Money will take between 9 and 11 working days, 90 percent of the time, i.e. span of 2 days.
- Francis' Bank-o-rama will take between 1 and 15 days, 90 percent of the time, i.e. span of 14 days.

Think of the problems that may occur by having the funds available early. The closing costs and first mortgage payment may begin before you had planned. What would happen if the funds are late? You may not be able to purchase the new house for the agreed price. You may have to carry the costs of the sale before you were ready. You may not be able to arrange time off from work at a moment's notice to sign the final documents.

If you are like most customers, you will find that having a definite time for a delivery or service is more important than the expected average time. It allows you to make your own plans and schedule things to happen at your convenience. Your customers are no different.

The real benefit of lean production is understanding and predicting changes in customer expectations and allowing you to react quickly. A full implementation of lean Six Sigma will drive the business culture toward one where everyone understands the demands of the customers, in terms of quality, features, and especially the timeliness of delivery. This understanding leads to being able to plan your capacity for meeting those demands. It is not necessary to be able to respond immediately to a customer demand, but being able to accurately predict delivery of service and communicate that to your customers is the key. Now the focus becomes internal, and management concentrates on making optimum use of resources to deliver the service or product. The continuing quest is for constant and never ending improvement.

1.4   Typical Problems in Transactional Projects

Information is the unit of work that travels throughout the transactional portions of an organization. It does not build up physically the way inventory can. It is difficult to assign a value to a unit of work the way you can for pieces of machinery or components awaiting assembly. It is also difficult to translate the effect of missing information at each stage on the overall throughput of the organization as a whole. As you are beginning your project you may have already found:

- It is difficult to define a *defect*, but the feeling is that it is very high.
- You have a lot of discrete data, but not a lot of continuous data.
- The financial impact and benefits of the project are difficult to quantify.
- Your data comes from a variety of different information systems.
- There is a tremendous amount of historical data and it is filled with errors.
- The business processes are widely distributed in location.
- "Inventory" or *work-in-process* (WIP) is not visible.
- The cycle time data is non-normal, making it difficult to calculate process capability and search for the "Vital X" using the set of statistical tools you have.
- The financial impact of WIP is unclear.
- Performance metrics of workers may be causing the problems you see and resulting in resistance to the project.
- Your organization spends a huge amount of time dealing with recurring problems, expediting orders, and placating customers.
- You have heard that improvement techniques, such as 5S and Kaizan, have achieved great benefits in other businesses, but you are unsure where would be the best place to apply them in your business.

1.5   Change and Corporate Culture

Every company has its culture. Your organization might be called stable. It is careful and slow to change with a general aversion to change. It could be called quick and nimble. Rapid changes are required to track constant shifts in competitive environment and customer needs. There is no best way for a company. Rapid change can be unnerving for employees and customers alike; slow bureaucratic organizations can be equally frustrating to deal with.

When Scott McNealy joined the board of directors at GE he commented that:

> Jack had developed a learning organization that can spin on a dime because he's got these black-belt, Green Beret-type folks infiltrated

throughout the organization. So when the word comes down this is the new initiative, away they go. (Fortune, *May 1, 2000*)

Six Sigma was embraced by GE as a clear, data driven methodology for attacking business problems. Six Sigma, lean manufacturing, or any other business initiative requires constant encouragement and guidance by the senior executive. Their endorsement must pervade the career advancement ladder, the hiring, and training programs. Any improvement program will be rooted in identifying and reporting business problems to the people who have the power to change the business processes that caused the problems to begin with.

If the corporate culture is to "kill the messenger," then the programs will suffer a quick and predictable death.

Piet Van Abeelen, the GE corporate VP of Six Sigma, saw how the variation in production output of raw Lexan caused financial and other problems, both internally and externally to the business, when he was running GE Plastics in Europe. It took an iron will on his part to tell production to shut down when they had reached their targets for the day. The remainder of the day was used to maintain the equipment, breaking the chain of machinery breakdowns causing shortfalls in production, and the resultant "catch-up" production causing subsequent breakdowns.

Business improvement programs and others like them take years to fully integrate into the corporate culture. The emphasis of lean and Six Sigma in this book is on the tools for execution of individual improvement projects and not on the wider problem of changing corporate culture.

2

Evolution of Lean Six Sigma, R-DMAIC-S

2.1 Six Sigma—DMAIC

The history of Six Sigma is well documented. In brief, it started at Motorola in the late 80s in order to address the company's chronic problems of meeting customer expectations in a cost-effective manner. Instead of thinking of quality as an inspection problem conducted after the fact, it was initiated at the front end of the process and continued throughout the manufacturing process. Each improvement project was organized into the four phases of:

Measure (M)—identify what your customers want or need and assess how you are failing to fulfill their expectations.

Analyze (A)—identify the internal causes of the problems.

Improve (I)—make changes to the product or service to improve it.

Control (C)—put signoffs or monitoring programs in place to ensure the improvements continue.

Each project was managed and reviewed at the conclusion of each phase.

Larry Bossidy, the CEO of AlliedSignal and ex-GE executive, brought success stories to the attention of Jack Welch, the CEO of GE. Jack took to the program completely and applied it across all of GE. Promotions were "frozen" throughout the company until everyone received training. When Jeff Immelt took over as CEO, in early September of 2001, he repeated GE's emphasis on using Six Sigma to achieve a company wide customer focus and individual career success.

Something that is impressive about the program at GE is how it continues to expand to all parts of the business where customer contact is made. Instead of being a program of "Inspected by No. 73", it evolved over the years to become an efficient system of business process improvement with customer focus and solid financial benefit to the company. Senior business leaders at GE must have received Six Sigma training and completed a number of projects before advancing in their careers.

Largely owing to the initial failure of the initial projects to deliver the expected financial impact, GE quickly added an extra phase to define and manage the improvement project. The *Define, Measure, Improve, Analyze, and Control* (DMAIC) structure has now become an accepted standard for Six Sigma project execution and management.

2.2 The Toyota Production System

The *Toyota Production System* (TPS) is at the heart of Toyota's manufacturing excellence. It pervades all aspects of the business in the same manner as Six Sigma does at GE. The TPS is commonly called *lean manufacturing* or simply *lean* by other industries, although lean is a subset of the TPS—an overall business philosophy.

In *The Toyota Way**, the TPS is summarized in fourteen principles:

1. Management decisions should be based on long-term philosophy even at the expense of short-term financial goals.
2. Create continuous process flow to bring problems into focus.
3. Use "pull" systems to avoid overproduction.
4. Level out the workflow.
5. Build a culture of stopping to fix problems and to eliminate rework.
6. Standardize tasks to facilitate predictability.
7. Use visual control systems to make problems visible.
8. Use reliable technology that serves your people and processes.
9. "Grow" leaders who understand the work and philosophy, and teach it to others.
10. Develop exceptional people and teams.
11. Respect your extended network of partners and suppliers by challenging and helping them.
12. Go and see the process yourself to thoroughly understand the situation.
13. Make careful, informed decisions slowly by consensus; implement rapidly.
14. Become a learning organization through reviews and continuous improvement.

---

*Jeffrey K. Liker, *The Toyota Way*, McGraw-Hill, NY, 2004.

Toyota's unique approach to manufacturing is the basis for much of the lean manufacturing movement. *The Machine That Changed the World* compared the philosophy of the manufacturing methods, worker incentive programs and job descriptions, reaction to customer needs, and continuous improvement typified by the automobile manufacturing industries around the world.[†] The manufacturing philosophies evolved from the craft production of customized products in Europe to the mass production of standardized products in North America to the lean production of high quality products by the Japanese.

The book *Lean Thinking,* addressed the question of how to achieve the results shown by Toyota.[‡] It showed a series of tools for improvement and general guidelines for setting up lean manufacturing environment. Their five step approach is to:

1. Specify value by specific product.
2. Identify and map the value stream for each product.
3. Make value flow without interruption.
4. Let the ultimate consumer "pull" value from the manufacturer.
5. Continuously pursue perfection.

2.3 Lean versus Six Sigma

We dealt with a small electronics manufacturing company that was in the process of implementing Six Sigma. The program was instituted by the parent company and they sponsored one person to attend Six Sigma training. The leadership attended a 1 to 2 day executive orientation session. The VP for quality was the single Six Sigma resource for the company.

The VP of operations had attended a conference on lean manufacturing and was very enthusiastic about applying the tools to solve his biggest problem—inventory turns were killing the company in a market where products can take two years to develop and would become obsolete quickly thereafter. He had identified that the practice of hanging onto inventory was having an impact on working capital and quickly became less valuable as time passed.

The argument that ensued was whether to use lean or Six Sigma tools to justify the decision. The two toolkits appeared to be mutually exclusive. The best

---

[†]James P. Womack, Daniel T. Jones and Daniel Roos, *The Machine That Changed The World,* Harper Perennial/Harper Collins, NY, 1990.
[‡]James P. Womack and Daniel T. Jones, *Lean Thinking,* Free Press/Simon & Shuster, NY, 2003.

problem solving techniques were fought by "non-invented by me," resistance on both sides. Both leaders tried to convince others in the company that their own methodology was superior to the other, and claimed credit for the project improvements. A large part of their solution was to merely sell obsolete inventory at a considerable discount. Neither of the business leaders identified that a disconnected communication process between product development, sales, and marketing was responsible for the problem in the first place.

2.4 Lean Six Sigma

A business process is a series of interconnected subprocesses. The tools and focus of Six Sigma is to fix processes. Lean concentrates on the intercon-nections between the processes (Fig. 2.1).

Defect reduction is a central theme in Six Sigma. The root causes of the defects are examined and improvement efforts are focused on those causes. The

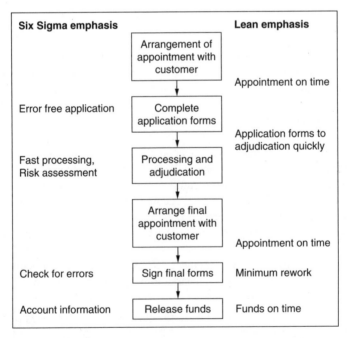

Figure 2.1   Emphasis of Six Sigma and Lean for Mortgage Processing

emphasis on defect reduction in lean is indirect in that defects can cause delays to occur. The combination of the two philosophies can be summed up as:

"Reduce the time it takes to deliver a defect-free product or service to the customer."

It is not a question of whether the customer wants it right *or* quickly, but both. The clock does not stop ticking until the customer is satisfied and has completed the transaction or received the product.

2.4.1 Lean Six Sigma—The R-DMAIC-S Cycle

The GE culture has always been one of aggressive growth guided by data driven decisions. The impact of management decisions on shareholder value is always considered. Individual performance goals are made clear to the employees, and Jack Welch's long-term strategy was well communicated within the company.

When this level of executive vision is less than perfectly clear to everyone in the company, improvement projects can lack the impact they might have. *Black Belts* (BBs) and *Master Black Belts* (MBBs) end up working on projects in familiar areas with a small overall impact on the company and the customers. When a strong credo of data-driven decision making is missing in your corporate culture you must have add two additional aspects to your DMAIC improvement projects. The two extra sections will help you to identify projects with high customer impact and will sustain the gains of adopting a lean Six Sigma culture. Your projects should be initiated and guided by a Recognize (R) stage and reported and evaluated as a part of the Sustain (S) phase (Fig. 2.2).

The complete project portfolio will follow the following R-DMAIC-S steps:

1. Recognize—Be aware of the need for change.
   - Identify the systematic problems or significant gaps in your business–What are the problems in the business?
   - Articulate the business strategy—What do the directors want the company to deliver?
   - Manage the effort—Does everyone in the company know what to expect of lean Six Sigma projects?

2. Define—Formulate the business problem.
   - Change your viewpoint—How do your customers look at you?
   - Develop the team charter—What are the team members expected to do?

Figure 2.2 R-DMAIC-S Cycle. Sustain and Recognize Phases Use the Data from Completed Projects, the Strategic Business Plan, and the Lean Six Sigma Deployment Plan

- Scope the business process—What subprocesses of the business are you going to work on (and not work on)?
- 3. Measure—Gather the correct data.
  - Select the *critical to quality* (CTQs)—What are you going to improve?
- Define the performance standards—What is the best way to measure?
- Validate the measurement system—Can you trust the output data?
- 4. Analyze—Change to a statistical problem.
  - Establish process capability—How good or bad are you today?
  - Define your performance objective—How good do you have to be?
  - Identify multiple business processes and sources of variation— What factors (*X*s) could make a difference?
- 5. Improve—Develop a practical solution.
  - Screen potential causes—What is the real root of the problem?
  - Establish the relationship between the factors (*X*s) and the output (*Y*)—How can you predict the output?
  - Choose the best improvement strategy—How big an impact has your best solution made on the CTQs?
- 6. Control—Implement the practical solution.
  - Validate the measurement system for the factors (*X*s)—Can you trust the input data?
  - Establish operating tolerances for the factors (*X*s)—How tight does the control have to be?
  - Implement process control and risk management—Does everyone know how to maintain the gain in the project?

(7.) Sustain—Retain the benefits of the program.
- Report and evaluate the lean Six Sigma projects—Have you shown the directors of the program how your project made an impact?
- Integrate lean Six Sigma into business systems—How does the lean Six Sigma program integrate with other business initiatives?
- Identify new opportunities—What do we do when new problems arise?

2.4.2 Roles in Lean Six Sigma R-DMAIC-S Projects

The roles and responsibilities are split into three groups:

1. The executive are responsible for identifying "global" problems in the business and the business environment. During the Recognize phase, they formulate the long-term strategy and communicate the goals of the company on a periodic basis, usually yearly. The financial goals, competitive positions, and long-term market movements are considered. They formulate external performance metrics and goals and turn those into a consistent set of internal performance metrics and goals. From their viewpoint, lean Six Sigma will appear as Recognize-lean Six Sigma projects - Sustain. (Fig. 2.3)

2. MBBs take these internal metrics and goals and drill into the business processes that have the biggest impact on those goals. They must coordinate the projects of the BBs to align with the yearly corporate goals and choose the best subset of improvement projects on which to concentrate their efforts. The final results of the DMAIC projects are reported to the executive

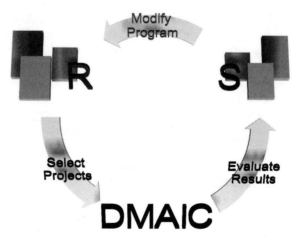

Figure 2.3   R-DMAIC-S Communication Plan

team on a periodic basis so that they may identify if the lean Six Sigma program is achieving the anticipated results. They will modify the metrics or program if it needs modification. From the MBBs' perspective, lean Six Sigma will appear as Recognize-DMAIC-Sustain.

3. BBs are the masters of DMAIC. They manage the project teams, collect and analyze data, and work with the project team to change the business processes. They finish a project by transferring the improved processes to the process owners. Project reports, milestones, and metric improvements are shared with the MBBs who consolidate them and report to the executive. The BBs will look at lean Six Sigma as R-Define-Measure-Analyze-Improve-Control-S.

2.5 The Product-Process Matrix

In *The Machine that Changed the World*, gave an excellent overview of the differences between the one extreme of craft production businesses that produce a small amount of customized product and the mass production businesses that create a large amount of product with very little variation.[§]

Businesses and business processes do not have to be either one or the other. Within a single business, new products or services start from individually specialized units, then the business processes evolve to become more efficient, streamlined, and high volume. The Product-Process matrix of Hayes and Wheelwright is a useful tool to illustrate the spectrum from craft production to continuous flow production (Fig. 2.4).[¶]

The horizontal axis charts the different types of products. On the right side of the figure are highly specialized products and services. These require highly flexible business processes that can be executed in a variety of ways depending on the needs of the customer. These types of processes are found in successful custom job shops and are well suited to the changing requirements of each job. This flexibility, however, comes at a cost to the business and the customer; customized solutions and services come at a premium, but may be the best solution for the market niche.

On the left side of the horizontal axis in Fig. 2.4, the products are usually a "one-size-fits-all" commodity where customers choose the cheapest

[§]James P. Womack, Daniel T. Jones and Daniel Roos, *The Machine That Changed The World*, Harper Perennial/Harper Collins, NY, 1990.
[¶]R.H. Hayes and S.C. Wheelwright, *Harvard Business Review*, 1979, pp. 133–140.

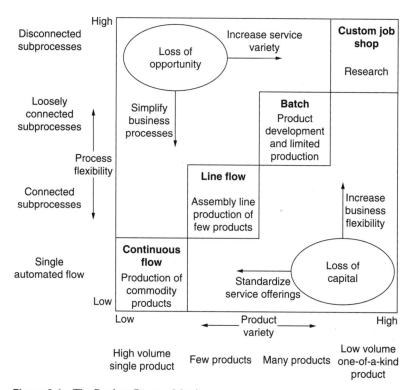

Figure 2.4 The Product-Process Matrix

product. In order to deliver this product, the business must be highly structured where economics of scale result in low unit costs. The business is inflexible and may offer only one kind of service, but this is highly specialized business with dedicated resources for single tasks, and is well suited for a high volume product.

Products that are produced with some variety, but still within medium to high volume require business flexibility that is intermediate to the two extremes. The diagonal is a continuum of business processes that have achieved the correct balance of standardization and flexibility to offer the ideal product or service mix. The off-diagonal regions are where there is a poor product-process match.

The lower right corner of Fig. 2.4 represents businesses that believe they must be highly specialized in available product offerings to satisfy customers, but have relatively inflexible and specialized business processes. Their present situation is that they have spent a large amount of cash in specialized resources and are not getting sufficient return on their investment. Their choice to improve the business is to:

- Standardize service offerings. Accept that their business processes are best set up for only a limited number of products or services, and they should concentrate on only those ones that can be produced with a combination of low margin and high volume. They will move to the left across the figure.
- Increase the flexibility of the internal business processes. Changes in customer needs must be addressed quickly and efficiently without having to redesign business processes with each service offering. They will move up the figure.

The upper left corner represents businesses that have failed to structure their business environment to be efficient in offering a limited set of services. Many business processes exist to make essentially a single product. The present situation is that significant opportunities have been lost because they failed to recognize that they had the flexibility to offer more variety to their customers. Their choice to improve the business is to:

- Increase the variety of service offerings. Accept that the business is capable of providing a customized service to the customer. They will move right across the figure.
- Simplify the business processes. Accept that their limited variety of product offering can be produced by a simplified subset of their existing business subprocesses. They will move down the figure.

A horizontal movement across the figure represents a marketing or sales opportunity, where the result is an increase in sales volume (to the left) or profit margin (to the right). The financial benefit is an increase in net revenue in either case. The impact on the customer is either a drop in price as sales volume increases, or an increase in variety of service offerings.

A vertical movement on the figure represents an internal change and involves costs to the business. Movement down the figure is the result of streamlining and standardizing business processes. Internal costs decrease when a complex customized business process requiring a dedicated team of corporate lawyers is replaced by three to four standardized services where contract specifications and terms have been worked out beforehand and require execution by less specialized resources. Movement up the diagram requires a degree of flexibility that may require more general resources.

Customers should be willing to pay more for specialized services. The impact on the customer is either a decrease in price and increase in speed as the business gets more streamlined (down) or an increase in flexibility as the business becomes less bureaucratic (up).

When we were presenting lean Six Sigma to a group of sales and marketing people at a large equipment manufacturing business, their objection was that all customers were different, all product offerings were different, the marketplace changes, and so on. In other words, they were trying to convince us that there was no standard process and that all sales are driven on the relationship between the salesperson and the customer. The folklore was that their business was highly specialized, requiring a larger, dedicated, and highly qualified engineering staff. Very long cycle times and high costs were the result of the highly specialized, customized products. They placed themselves in the upper right of the Product-Process matrix, in the "custom job shop" block.

When we started mapping the sales process, we found that the flow for existing sales was common in many aspects. For example, all orders required a detailed quote outlining specifications on price, technical performance, delivery, acceptance criteria, and payment. Each order was different, of course, but each one contained the same type of information. Some changes in subprocesses were accepted. Orders larger than a certain cutoff required a greater degree of financial scrutiny before a quote was sent to the customer.

Not every query from a potential customer required a detailed quote. The sales department was requesting a detailed quote for each potential order. These required weeks of engineering time that was not always necessary. The front end of the sales process was standardized by first asking the customer a number of screening questions and identifying the best and most timely response. A majority of potential orders were addressed by a sales representative who responded quickly by visiting the customer site with a basic product information booklet. The sales staff was trained to gather data on existing equipment, timeline requirements, and technical requirements and pass those onto the engineering and sourcing groups when a detailed quote was required. The completeness of the information decreased the cycle time for quote generation.

The projects we developed with the sales and marketing teams were centered around the front end of the sales process to provide the most appropriate information to the customer when they expected it. The back end of the sales process and order management had much less of an impact on the customer.

We worked with another business specializing in customized electronic control and communication equipment where each order was considered to be a customized solution. As in the previous example, this required each order to be managed by a product engineer. When we examined all orders shipped over the previous two years, we found that a majority of the orders could be handled by three or four common configurations with minor modifications.

We broke out these product lines and handled them as standard models allowing streamlining of parts sourcing, dedicated assembly line, and final testing. Cycle times on order-to-shipment dropped from weeks to days.

These examples illustrate a general trend—business processes changes lag behind business opportunity changes. As businesses grow, they will find themselves moving off the diagonal and into the upper left corner of the Product-Process matrix.

2.6 An Evolution of Service Offerings

Public companies are expected to provide to the shareholders three things.

1. Shareholders should receive a better return on their investment in your company than they could from depositing the same amount of money in a bank and collecting interest. Your company must add enough value to the products and services that the return on investment (dividends) is even greater. Your business must constantly control and reduce costs associated with production of the services it provides. This means that there will be constant pressure from the stakeholders to move down the Product-Process matrix.

2. Shareholders also expect year-to-year growth in revenue from sales. This is comparable to the increase in principal that bank customers enjoy. This expectation will result in pressure to move from one-of-a-kind products to higher and higher volumes of more standardized products.

3. Success in the market place is measured by market share. A high market share must mean that customers are willing to pay for your service given the array of choices from your competitors.

Think of some of the more successful service offerings that your business provides. It is likely that the service started off as a specialized, custom job offering. It was first produced with a "design-as-you-go" approach to the business processes evolved. As the service became more accepted, the business processes became more streamlined and automated, with lower costs and resulting in higher sales volumes. Different businesses will have varying numbers and timelines for the evolution and lifecycle of service offerings. A successful company will always have a mixture of service offerings in different stages of maturity with a constant flow of services from the upper right to the lower left corner of the Product-Process matrix.

2.6.1 A Static Mixture of Service Offerings

A full service hospital will have variety of service offerings that range from relatively standardized, low cost services to flexible, high cost ones.

Consider operating a world class service for age related cataract operations. This is a service where processes can be highly standardized and costs are fairly constant and predictable. A single ward or even an entire facility can be staffed with dedicated staff with standardized, specialized diagnostic and monitoring equipment, surgical tools, and procedures. Staff and patients can be scheduled and managed for maximum utilization of staff while providing reliable and high quality service to patients on an out-patient basis. A continuous drive to improve the service is possible given the repetitive nature of the work. Cost management is the key to continued success of the facility.

Now consider the opposite end of the spectrum—full service emergency. This facility must be managed to be able to respond to rapidly changing and unpredictable situations that run the full range of health related issues. The responsiveness comes at the detriment of cost-effectiveness. It would be ludicrous to expect an emergency department to have the capacity to schedule appointments to maximize the utilization of the staff members and physical facilities. Maximum flexibility is essential for the continued success of the emergency facility (Fig. 2.5).

If a general hospital attempts to have a single set of business processes to serve two distinct services, then it is doomed to either enormous cost overruns or customer dissatisfaction, perhaps with grave consequences. The administration of the general hospital has the daunting task of managing a static Product-Process matrix. Budget and staffing requirements will seem grossly unfair to some of the stakeholders in the hospital. Balancing costs and standardizing business processes across all patients' services will result

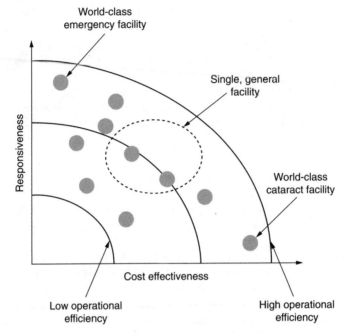

Figure 2.5   The Relationship Between Cost-Effectiveness and Responsiveness

in lost opportunity in the high volume, low margin, standardized, outpatient services, and lost capital in the low volume, high margin, highly responsive, emergency services. The best way to design the different service offerings is to equalize all aspects of risk across all different potential patients.

3

Recognize

3.1 Variation

"Variation is evil in any customer-touching process"
   Letter to Shareholders, General Electric Company 1998 Annual Report

A survey of major customers at a GE business showed that customers were not as concerned about speed as they were about predictability. The overwhelming response to the survey was not that GE should provide things quickly, but rather that the customers merely wanted things to occur when they expected them. Piet van Abeelen, the VP Six Sigma at GE Corporate pushed the idea that minimizing the difference between a promise date and delivery date was more important than reducing the average delivery cycle time. Before he moved to his position at GE Corporate, he was in charge of six GE Plastics plants in Europe. The plants manufactured raw Lexan beads and shipped them to customers on a daily basis for final fabrication. He identified that inconsistent production output from his plants created numerous problems for his customers. The simple business metric was not a sigma level or the number of defects per million opportunities, but *span*—the interval that would describe 90 percent of the process output. An example could be, "90 percent of our deliveries are between 5 days early to 8 days late for a span of 13 days."

The results of the improvements driven and measured by reducing the span on delivery resulted in an overall production increase of about 17 percent for each of the six plastics production facilities. By focusing on reducing inconsistent production in Europe, van Abeelen essentially gave GE Plastics the increased production of one extra plastics plant without having to construct it. The customers were delighted with the consistent supply, allowing better capacity planning with consequent benefits on their part. The success of the project was felt on the part of the GE shareholders and GE customers. When improvement is done well, then both parties can benefit simultaneously.

This example shows that you should concentrate on the customers' needs first, and the financial results will follow. Although it was not called lean, the *variance based thinking* (VBT) initiative had similar objectives of reducing variation in supply and demand to make for facile accurate prediction of plant capacity.

3.2 Identify the Business Needs—Strategic Planning

The reduction of variation in supplying product to the customer was only one aspect of a much larger strategic plan for the GE. It is not a panacea for all businesses. It will be one part of the strategic planning process for the business.

Ideas for individual projects may come from customer surveys, systematic problems that have failed all previous attempts at a solution, a changing marketplace due to changes in technology, customer wants, or entry of a competitor. These may give you a large number of good things to work on, but you should be able to explain the real problems of the business in terms that everyone in the organization will understand. You must be able to translate problems into a business plan with quantifiable benefits.

Your company will have a group responsible for strategic planning. They will usually be at a senior executive level and meet periodically to assess the market, business capacity, and customer needs; then conduct gap analyses to identify areas of improvement and report the results to the board of directors. In some businesses, strategic planning may be restricted only to new product development or marketing.

There is, of course, an enormous amount of information available on the process of business strategy development and financial analysis. If you are fortunate, then much of this information is available to you from your business leaders. Chapter 2 outlined the roles and responsibilities in the different *Recognize-Define, Measure, Analyze, Improve, Control-Sustain* (R-DMAIC-S) phases and showed that the business leaders are chiefly responsible for identifying the needs of the business. You should be able to talk with them and understand the language. This chapter is intended to give you an overview of what is involved.

3.3 Stakeholders—Shareholders and Customers

There are two groups of stakeholders for any publicly traded company. A successful business is one that understands and meets the needs of both groups.

The first group of customers is the company's shareholders. They are mostly focused on the financial aspects of the business. In order to satisfy them, your company must be able to give them a greater return on their investment than a commercial bank. They are focused on market share, growth, and profit.

The second group is the customers paying for your business' services or products. A common mistake is to assume that this group makes all decisions based on cost. Customers make decisions based on value, not cost. If that was not the case, everyone would drive the same model of automobile, deal with the same bank, and have the same cell phone plan. Clearly, that is not true. Every customer will assimilate an enormous amount of information before making a purchase. They are focused on quality and value.

If you are good at understanding and satisfying your paying customers, then success with the shareholders will follow. The tools outlined below will show how the two groups are related.

3.4 The Strategy of a Successful Company

In 1999 a *Fortune* cover story concluded that the ability to execute strategy was a key pillar of successful CEOs and a source of failure of poor ones. An earlier *Fortune* study showed that 70 to 90 percent of strategy failed owing to lack of execution.

The Strategy-Focused Organization,[*] showed a series of steps between the philosophy of the company—"Why do we exist?"—and personal objectives—"What do I need to do?"(Fig. 3.1)

Our focus will be to start you from an intermediate stage below the mission and vision statements. It is assumed that the executive and corporate directors have established about the first three steps of Fig. 3.1, and you are responsible for coming up with an execution plan.

The corporate strategic goals must be clear at the personal level in order that specific areas to focus improvement efforts can be identified. The communication must also provide a "route to the top" to show the effect of individual effort on corporate goals.

Although Fig. 3.1 shows a one-way flow from corporate mission to the strategic outcomes, it is never as clear as implied here. Lean Six Sigma

---

[*]Robert S. Kaplan and David P. Norton, *The Strategy-Focused Organization*, Harvard Business School Press, Boston, 2001.

Figure 3.1   Translating a Mission into Desired Outcomes

Master Black Belts (MBBs) and *Black Belts* (BBs) will be constantly required to define the downward flow of requirements and quantify the upward flow of capability.

3.5   The Balanced Scorecard

When Kaplan and Norton developed the Balanced Scorecard, they did it to align business activities to the strategy and monitor performance against strategic goals over time. A company's vision and strategy is developed and examined from four perspectives (Fig. 3.2).

3.5.1   Financial

How does the strategy appear to your shareholders with respect to growth, profitability, and risk management?

In the past there has been perhaps too great an emphasis on financial success as the sole measure of a successful company. The financial measures are the

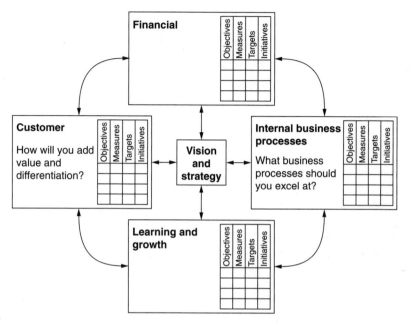

Figure 3.2   The Balanced Scorecard

final outcomes of determining the needs of the customers, designing a product or service, and finally marketing and selling that product or service in a cost-effective and competitive manner. The financial metrics are lagging indicators and short term measures that depend on the other three perspectives of the Balanced Scorecard. The traditional financial perspective may also include *return on investment* (ROI) analysis of new ventures and services, risk assessment, analysis, and mitigation or remediation countermeasures.

3.5.2 Customer

How does the strategy appear to your customers with respect to creating value and differentiation?

In businesses it can become easy to focus on day-to-day operations with the consequent interval perspective. If customers are not satisfied, they will eventually explore similar services performed by your competitors. Poor customer satisfaction is the leading indicator of poor financial performance in the future.

In the private sector, customers are the individuals or organizations who pay for services or products. In the public sector and government, customers could be taxpayers, representatives in government, or managers of government portfolios. In most publicly traded businesses the customers are both the ultimate person paying for the service or product and the company's shareholders and board members.

3.5.3 Internal Business Processes

How does the strategy appear with respect to the internal business processes you employ on a day-to-day basis?

There are two main streams of processes internal to businesses. The value stream is directly associated with the generation of revenue and involves the generation of the service or product. The second stream is composed of the support functions of the business and is usually associated with cost.

When Jeff Immelt became CEO of GE, he instituted his "front office-back office" initiative. In order to satisfy the customers he wanted to increase customer touch time (front office) while decreasing the cost involved (back office). This single statement had implications to the customers in that they would enjoy greater time with the product and sales representatives of the company and less time with quality assurance, accounts payable, and shipping departments. The shareholders would also benefit from better financial performance of GE as a whole. Immelt went on to point out how Six Sigma was going to be used to change business processes to achieve this goal.

3.5.4 Learning and Growth

To achieve your vision, how will you sustain your ability to change and grow?

Learning is much more than training. It includes the ability of employees and the organization as a whole to continually explore novel solutions to new problems. This continues as a constant cycle of business and operational planning. The VP lean Six Sigma, MBBs, and human resources personnel should incorporate the Learning and Growth *quality function deployment* (QFD) in succession planning, BB selection, and evaluation. Yearly program planning should center around the evaluation of the effectiveness of the lean Six Sigma implementation.

Within each of the four perspectives each objective has a measurement, a goal, and a specific initiative. This decomposition is comparable to the phases in a Six Sigma project (Fig. 3.3). Note that the Control phase of lean Six Sigma does not have an explicit corresponding component in the Balanced Scorecard.

The presentation of the Balance Scorecard in Fig. 3.2 implies that the four perspectives are independent ways of looking at the vision and strategy. The relationships between the four perspectives are shown in the figure as double-headed arrows, but not explicitly examined or demonstrated.

Lean Six Sigma projects must show cause and effect relationships with an impact on corporate strategic goals. This must include performance measurements, targets and goals, timelines and tollgates, competitive information, and clearly highlight the most important areas for improvement efforts.

Balanced scorecard	Six Sigma phase	Operations example	Banking example
Perspective	–Project definition	–Improve operational excellence	–Reduce cost per customer
Objectives	–Voice of the customer –Competitive benchmarking –Strategic goals	–Improve inventory management	–Migrate bank customers to use on-line services
Measures	–Metrics design –Data collection plan	–Inventory turns –Inventory net present value	–Number of on-line transactions –Value of on-line transactions
Initiatives	–Improve –Process benchmarking	–Product, order and capacity management –Product lifetime management	–Develop applications, internal and customer training, telephone and on-line help
	–Control and monitoring	–Order, inventory and manufacturing levels	–Customer surveys and weekly transaction reports

Figure 3.3   The Balanced Scorecard and Lean Six Sigma

We explicitly incorporate knowledge at all levels of the organization and articulate the outcomes from the strategic planning process by sequencing the four perspectives in a cascade of "what to do" and "how to do it" relationships.

3.6 The Strategic Planning QFD

It is important for everyone in the business to understand the cause and effect relationships between the four perspectives of the Balanced Scorecard. The usual implementation is simply a map showing untested assumptions. If the implementation is not done in a comprehensive manner and incorporated into the strategic planning process, the result may be nothing more than a *key performance indicator* (KPI) scorecard.

In order to introduce some statistical rigor, data derived from customer and employee surveys, market intelligence, competitive benchmarking, and financial analysis are incorporated to focus improvement efforts. Statistical hypothesis testing may be required to establish these relationships.

The QFD is an excellent tool to combine these multiple sets of quantitative data and to quantify the cause-and-effect relationships between corporate goals and individual process improvement efforts. The layout of the tool shows the links between the needs and goals, and the actions required. The layout is shown in Fig. 3.4.

In parallel to the four perspectives of the Balanced Scorecard, one QFD matrix is constructed for each one of the financial, customer, internal business processes, and learning and growth categories. The result is a series of solid cause-and-effect relationships linking the financial impact of corporate strategy to the needs of the customers and individual improvement projects (Fig. 3.5).

One difference between the Balanced Scorecard and the Strategic Planning QFD is that the upper QFD matrix includes some of the components of both the strategy and the corporate vision. For example, it should include the traditional strategic goals of revenue and profit targets, as well as the less tangible vision targets such as diversity, respect for the environment, and corporate responsibility and transparency. The metrics of the upper, Corporate QFD must include *everything* the company is planning to deliver, either in the short- or long-term. We once worked with a small manufacturing business to help select lean Six Sigma projects after the initial training and implementation plan for a "pilot" program were completed

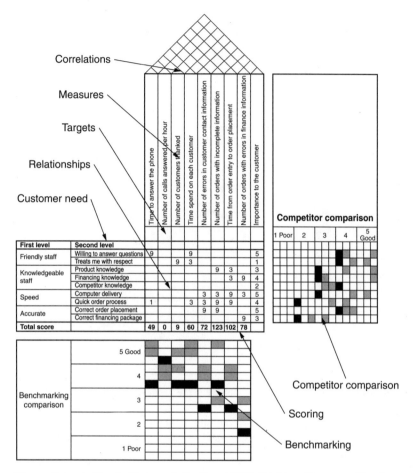

Figure 3.4   A Single House of Quality (QFD) Linking Customer Needs to Measurements with Three Vendors

by another vendor. The requirements for the engagement are commonly requested; they wanted a small number of high-impact projects that were easy to execute to prove the concept of lean Six Sigma to justify a larger initiative.

The biggest problem was that metrics for the lean Six Sigma implementation were managed separately from other corporative initiatives. By placing lean Six Sigma goals and metrics outside other business initiatives, senior managers would dilute lean Six Sigma improvement efforts to focus on their

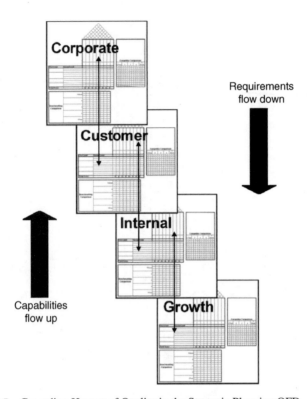

Figure 3.5   Cascading Houses of Quality in the Strategic Planning QFD

own improvement projects. The resultant competition for business resources resulted in duplication of effort, project resistance, and subsequent failure of the program.

Diavik Diamond Mines Inc. (DDMI), based in Yellowknife, Northwest Territories, Canada, is a subsidiary of Rio Tinto plc of London, England. The Diavik diamond mine in Yellowknife, is an unincorporated joint venture between DDMI (60 percent) and Aber Diamond Mines Ltd. (40 percent), a wholly owned subsidiary of Aber Diamond Corporation of Toronto, Ontario. The implementation of business practices for the Rio Tinto group of companies is defined in the corporate document, *The Way We Work*, and includes social responsibility as a central tenet.

Rio Tinto's elements of social responsibility such as safety, occupational health, and the environment would be included in the Corporate QFD alongside the financial goals. Improvement efforts in all segments of the business would now roll up into the Corporate QFD. Improvement projects aimed at quickly delivering effective safety training to new employees will then show an impact on the workplace safety and financial targets in the corporate strategic plan.

Starting at the corporate perspective, the corporate goals flow down to give directions to individual improvement projects. The improvement in capability at the individual improvement project level flows up to make an impact on corporate goals. Automatic data feeds from *enterprise resource planning* (ERP) systems allow the design of a corporate dashboard for constant monitoring of the results of improvement efforts and programs.

Once the QFDs are complete and communicated within the company, employees have the means to see how they can make an impact on the corporate goals.

3.7 Handling Different Business Units

When a business has a number of distinct parts, the consolidated financial statements may be broken into the constituent parts of the organization. This organization of business units is in recognition that each operating unit has its own separate group of customers and *critical to quality* (CTQ) characteristics and operational risk factors. Each business unit can have its own independent or overlapping financial and corporate goals, but they will become consolidated in the Corporate QFD.

Scotiabank, for example, has separate sections for each of the three major business units:

1. Domestic banking—retail banking, wealth management, small business, commercial banking
2. International banking—from operations outside Canada
3. Scotia capital—investment and corporate banking

Each business unit should have its own Balanced Scorecard QFD for customer and internal processes to facilitate project selection and the calculation of benefits (Fig. 3.6). Fluctuating foreign exchange rates, for example, will be a greater factor for the international business unit and lesser for the other two units.

Portions of the Learning and Growth QFD may be shared by the human resources departments of each business unit, and should be coordinated by

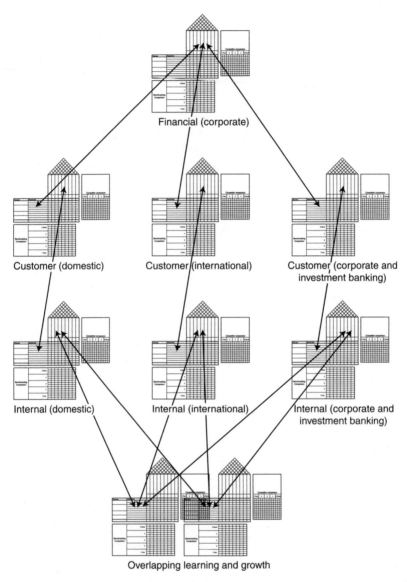

Figure 3.6   Cascading QFDs for Separate Business Units in a Bank

MBBs or the VP lean Six Sigma at the corporate level. This is sometimes referred to as *sharing best practices* between business units. It is likely that there will be overlapping areas for project focus and all three groups should communicate for good operational definitions for commonly used terms, such as operational risk or revenue.

Within each business unit, the different parts of the organization will be considered within the internal QFDs.

3.8 Finance Must be Part of the Process

There is a strong financial benefit component of lean Six Sigma and the reporting must be done correctly. When projects are reported with fluffy benefits, then the entire program loses any legitimacy. It is important that the upwards reporting of financial metric improvement is done in a clear and rigorous manner. It is suggested that a financial representative be a part of each lean Six Sigma team for a project with a projected financial benefit of over $50,000, and

- Be responsible for assigning the category of expected financial benefit at the Define report stage. (Section 3.19, "Financial Benefits Buckets")
- Review the financial benefit at the end of the Improve stage.
- Audit each project for continued financial benefit one year after implementation.

3.9 Reporting the Benefits—The Corporate Dashboard

It would be convenient to choose one or two metrics to monitor and judge whether the company is doing well. It may be possible as an external observer to look at growth, profitability, and market share, but this information is too coarse and usually too slow to use as a day-to-day insight into the effectiveness of business initiatives. Corporate dashboards can be designed to constantly report the performance of the company against targets for any of the metrics in the Strategic Planning QFD. The dashboard must contain at least the metrics in the Financial QFD, some of the ones in the Customer QFD, and a few of the important ones from the Internal and Learning and Growth QFDs. The metrics should be consistent across all business units and complete in their strategic scope.

The number of metrics and the individual detail will be greater than the few examined by external analysts. The external measurement of pure profitability will be broken into smaller segments for the executive dashboard.

Consistent financial summaries broken out by geographical sales regions, product groupings, product mixture, product/service vitality, numbers of improvement projects in different business units, and number of BBs trained, customer attrition and satisfaction are only a few examples of metrics suitable for a corporate dashboard.

Great care should be taken when choosing metrics for the dashboard. We have consulted with many companies that have tried to drive a culture of customer focused continuous improvement, when everyone in the company saw the heads of business units who consistently met only the targets on revenue and profitability get large bonuses and executive accolades.

We were conducting some Six Sigma consulting with a small business unit that had been recently acquired by GE. The manufacturing group was trying to provide services as quickly as possible while the sales group was selling as quickly as possible, yet the business was not profitable. Everyone was working flat out and trying to come up with ways of cutting costs or increasing productivity. We asked how the sales group were compensated and found that they all had sales revenue targets. We also found out that the proposal hit rate was nearly 100 percent.

A quick benchmarking exercise showed that nearly every proposal was accepted because they were priced very low. The sales team still received bonuses, while the problem of driving down costs was now in the hands of manufacturing. The sales team was somewhat aware of the effects on the company, but concentrated on their own metrics. The result was that the sales team was selling manufacturing capacity below cost and driving the company faster and faster into the ground. The sales team laughed when someone said, "That's OK, we'll make it up in volume."

Finance reviewed the previous contracts, human resources had to negotiate and redesign the incentive and compensation package, and some customers were reluctant to accept the new pricing guidelines. Since they previously believed they were the only group in the company that generated income, the sales team was not happy that they thought they might be labeled as the cause of the downfall of the company when they were working hard on delivering and exceeding their sales targets.

There was enormous resistance when we told the sales staff that they would have to turn away business if it did not exceed a minimum margin target. The sales team now had to conduct a commercial review of each new contract above a particular dollar cutoff on a weekly basis. The finance team member would reject those that did not pass the margin cutoffs. The CEO showed the impact of these new policies on the business as a whole to address the resistance, and the business unit was turned around quickly. Manufacturing, the original focus of the improvement efforts, was left unchanged and was more than capable of handling the workload.

Carlos Gutierrez, when he was CEO at Kellogg, was credited with turning around the cereal giant since 1999. The company had previously tried to maintain unrealistic earnings-per share targets, skimping on *research and development* (R&D), and marketing while concentrating on shipping large volumes of product. In part he:

- Established clear metrics so that everyone in the company knew how his job contributed to meet corporate targets.
- Changed metrics from volume of product to dollars and margin.
- Grew sales by shifting production and resources to high margin products.
- Used profits to fund more R&D to develop higher margin products.
- Overhauled daily tracking systems to record dollar sales, not weight, and altered bonus plans to reward profits and cash flow, not volume.

3.10 Incentives, Compensation, and Certification

The visibility of corporate metrics is only half the story in making lean Six Sigma successful. During the 1997 annual meeting of all GE executive in Boca Raton, a memo from Jack Welch (Chairman and CEO), Paolo Fresco, and John Opie (Vice Chairmen and Executive Officers) announced that all executive band promotions were "frozen" until candidates had received Green or BB training. Jeff Immelt made a similar announcement when he took over as GE's CEO in September 2001.

The examples in the previous section should emphasize the importance of correctly aligning compensation programs with corporate goals. This issue should be addressed as part of the Learning and Growth QFD during strategic planning (Fig. 3.5).

Master Black Belts and quality leaders should have about 25 percent of their compensation tied to delivering lean Six Sigma benefits. Black Belts should have about 10 percent tied to benefits. This is likely to be a sensitive issue and requires the senior executives to work together to make a well planned and accountable system for quantifying benefits. The Strategic QFD will be the primary source of the metrics.

The credit for project benefits may be either spread between the team members or retained by only the project leader. This is a complex choice— each option has potential pluses and minuses. There is no solution that will work in all situations and all companies, but it is absolutely required that this is clearly defined for success of the program.

It will take about 18 months to 2 years for a BB to learn the practical application of the tools and to manage the lean Six Sigma teams. In a lean Six Sigma implementation, the first wave of BBs should have the necessary experience by the time the external lean Six Sigma provider has finished the engagement. Human Resources should have integrated experienced MBBs and BBs into the training delivery and determined the criteria for certification of the candidates as they are being moved back into operational roles. This group of graduates should assist in the ongoing refinement of certification guidelines. Certification should include:

- Completion of training
- Successful execution of several projects with lasting benefits and customer impact
- Mentoring and training other team members
- Mentoring and training other BBs
- Technical proficiency as measured by a written examination or interview

3.11 Elementary Financial Analyses

Some relationships between input and output variables in the financial section of the Strategic Project Planning QFD are defined by convention. This is especially true in the analysis of data reported in the financial statements at the corporate level. For example, the relationship between cost of goods sold and ROI is established by *Generally Accepted Accounting Principles* (GAAP).

A complete and detailed financial analysis of the performance of the company is constantly being conducted by the internal financial team. Their analysis is used by the executive to monitor and guide the performance of the company and to assess the impact of corporate strategy. Their input will be essential in the Financial QFD section. They must explicitly provide the financial goals of the company. The links in the Financial QFD will provide the final step for quantifying the financial benefits of the lean Six Sigma efforts during project closeout.

3.12 General Purpose Financial Statements

There are two important points about lean Six Sigma

1. The corporate goals of the company must clearly flow down from the Corporate QFD, such that project selection information can be gleaned directly from the internal process QFD.
2. Projects that result in a reduction in cycle time can have an impact on many different financial indicators.

Special interest groups, such as government regulatory bodies, may require specialized financial reports, but most analysts and external users of accounting and financial information will use the following to judge the financial health of the company (Fig. 3.7).

1. Income statement (statement of earnings, profit, and loss statement)–states the results of the operations of the company over a period of time. Examples of revenues and expenses include:
   - Sales
   - Cost of Sales
   - Expenses
2. Balance sheet–states the investments (assets) of a business and the financing of those investments (liabilities and equity) at a particular point in time. Key accounts are:
   - Assets–receivables, inventory, plant, and equipment
   - Liabilities–accounts payable

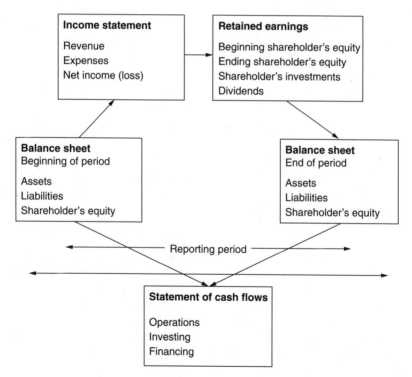

Figure 3.7   Components and Relationships Between Financial Statements

3. Statement of retained earnings–enumerates the portion of the shareholder's equity that has been earned, but not paid out to the shareholders as dividends. This stays in the business for future operations.
4. Statement of cash flows with notes related to all four statements–changes in accounting procedures, and major changes to the structure of the business through acquisition or divestment will have effects on all statements.

3.13 Cash Flow Reporting

The purpose of the statement of cash flows (Fig. 3.7) is to report how a company obtains its cash, where it spends it, and what is the change in cash from one reporting period to the next. All transactions of cash inflow and outflow are classified into cash from:

- Operating activities–the normal operation of the business including items such as cash and credit sales, refunds, wages, taxes and services, and payments to suppliers of raw goods.
- Investing activities–cash not required for the operation of the business may be involved to sell or purchase debt investments, to make loans to noncustomers, purchase or sell equity investments, or pay for future contracts.
- Financing activities–these are defined as those transactions that involve the company's owners and creditors. Activities will include issuing shares, paying dividends, covering withdrawals by the owners, and issuing bonds and notes.

External investors will look more favorably on companies that generate more cash from operations rather than selling capital assets. Lean Six Sigma will affect mostly the cash flow from operations.

This form of the cash flow statement is quite coarse in that it is only the summary of a year's worth of activity. It will not distinguish between a company that bills all customers on January 1 and collects all invoices on December 31 from another company that bills and collects throughout the year. A much more detailed analysis of cash flow for lean Six Sigma projects will be done during the DMAIC portion of the projects using what the accounting practice refers to as the *Direct Method*, where all individual transactions will be used to calculate the true nature of cash flow.

3.14   Competitive Benchmarking

At the macroscopic level, quick preliminary analyses can result in shining spotlights on selected areas of the company that require more work. This may result in showing that 80 percent of the company's operations are doing well and focuses the resources on the remaining 20 percent.

Benchmarking is one tool that can be used to make comparisons against:

- Internal divisions–compare sales regions or market segments.
- Competitors–address market share issues.
- Industry standards–banks, in general, for example.
- Heuristic guidelines–return on investment should be better than the open market interest rate.

Whether your revenues are larger than your competitors, or whether your expenses smaller than your competitors, they are simple comparisons of the performance of the company in the face of competition. These comparative benchmarks are not very helpful in the diagnosis of the operations of the company, but may be indicators requiring a more in-depth analysis.

3.15   Horizontal and Vertical Financial Reporting

You may be required to do an in-depth analysis of specific areas of your business with a more rigorous focus than benchmarking. The financial "guru" on the team should help in finding and analyzing the financial data broken out by divisions or sales regions.

Horizontal analysis is a comparison of a company's financial condition and performance across time. Horizontal analysis can reveal changes in market trends and is usually a part of the corporate dashboard. A summary of the last six months of monthly sales data is usually enough to identify non-seasonal trends. A common example of horizontal analysis is the comparative balance sheet or comparative income statement seen in annual reports. It is customary to report absolute changes in addition to percentage changes. The comparative income statement for SoftCo in Fig. 3.8 shows increase in nearly all indicators as a reflection of the tremendous growth of the company in the past year. Even though there is a large percentage increase in loss from foreign exchange, the absolute amount shows that it remains as the smallest expense. The cost of sales has increased in 2005, but so has the gross income from sales (Fig. 3.8).

SoftCo Inc.
Comparative income statement
For years ending Jul 31, 2004 and 2005

(in thousands)	2005	2004	Amount increase (decrease) in 2005	Percent increase (decrease) in 2005
Sales	167,907	108,333	59,574	55%
Cost of goods sold	37,103	25,419	11,683	46%
Gross profit	130,805	82,914	47,891	58%
Expenses				
Advertising	35,480	20,764	14,717	71%
Administration, selling and general	31,356	18,599	12,756	69%
R&D	11,095	6,470	4,625	71%
Amortization	6,261	4,236	2,026	48%
Foreign exchange loss (gain)	1,556	(490)	2,046	417%
Total expenses	$85,748	$49,578	$36,170	73%
Income from continuing operations	$44,910	$33,005	$11,905	36%
Interest income	2,986	2,481	506	20%
Income from continuing operations before income taxes	$47,896	$35,486	$12,411	35%
Income taxes	15,054	14,336	718	5%
Net income	$32,843	$21,150	$11,693	55%

Figure 3.8 Horizontal Analyses Using a Comparative Income Statement

Vertical analysis is also called *common size analysis*. It is usually easier to compare different line items from year to year by scaling all the numbers to a percentage of a base amount. The comparative income statement for SoftCo from Fig. 3.8 has been reformatted such that items are expressed as a percentage of total revenue for each year. The common size analysis shows that the company has not really radically changed the proportion of income or expenses as it has grown larger. The vertical analysis of Fig. 3.9 shows more easily that the cost of goods sold has decreased slightly as a percentage of sales (23.5 to 22.1 percent) even though the company has seen tremendous growth (Fig. 3.9).

3.16 Ratio Analyses

External users are primarily interested in the company's financial health as measured by:

1. Liquidity and Efficiency–the ability to meet short-term obligations and to effectively generate revenues. A few of the ratios in this group may help in identifying business process areas for improvement projects.

SoftCo Inc.
Common-size comparative income statement
For years ending Jul 31, 2004 and 2005

(in thousands)	2005	2004	Common-size percentages	
			2005	2004
Sales	167,907	108,333	100.0%	100.0%
Cost of goods sold	37,103	25,419	22.1%	23.5%
Gross profit	130,805	82,914	77.9%	76.5%
Expenses				
Advertising	35,480	20,764	21.1%	19.2%
Administration, selling and general	31,356	18,599	18.7%	17.2%
R&D	11,095	6,470	6.6%	6.0%
Amortization	6,261	4,236	3.7%	3.9%
Foreign exchange loss (gain)	1,556	(490)	0.9%	(−0.5)%
Total expenses	$85,748	$49,578	51.1%	45.8%
Income from continuing operations	$44,910	$33,005	26.7%	30.5%
Interest income	2,986	2,481	1.8%	2.3%
Income from continuing operations before income taxes	$47,896	$35,486	28.5%	32.8%
Income taxes	15,054	14,336	9.0%	13.2%
Net income	$32,843	$21,150	19.6%	19.5%

Figure 3.9   Vertical Analyses Using a Comparative Income Statement

2. Solvency–ability to meet long-term obligations and to effectively generate future revenues. These ratios tend to examine very long-term financing and have less to do with day-to-day operations than those for liquidity and efficiency.
3. Profitability–ability to produce rewards for the shareholders (investors) beyond the return from the market interest rate. The liquidity of the business concentrates on turning outgoing cash into incoming cash quickly, whereas profitability is the ability to turn outgoing cash into proportionately more incoming cash. The ratios discussed in this group should show that an increase in the numerator of a fraction has the same effect as a decrease in the denominator. It seems that management is usually more concerned with "bottom line" savings than with "top line" increase in revenue. They should be interested in both to increase profitability.
4. Market–ability to generate positive market expectations. For public companies, the market refers to the people or institutions that purchase or sell stock in the company based on their perception of value and risk.

External analysts use a number of indicators to assess the company in each of these areas while comparing them to other market leaders or market segments. We will review some of them here, but will limit ourselves to ones that are more likely to be influenced by operational and transactional lean Six Sigma projects than by changes in corporate financing or share structure.

As you begin your project, examine the different ratios discussed here and consider the effects that your potential improvement project may have on all the terms in the ratios. You may also see if you can identify projects based on the different components of the ratios. Discuss the financial implications with the financial representative on your team.

As you end your project, you will have to quantify the financial benefits of the project, once again, and discuss this with the financial person on your team.

3.17   Financial Liquidity and Efficiency Analysis

Efficiency in this context refers to how well a company uses its assets; *liquidity* is a measure of how well a company meets its short-term cash requirements. Liquidity is affected by the timing of cash inflows and outflows (cash flow). Transactional inefficiencies in manufacturing, billing, and collections can lead to problems with liquidity. Financial analysts measure it because it tends to portend lower profitability and inability to execute contracts.

Profitability is the primary goal of most managers, but a company cannot maintain itself without careful management of cash. Since payments made to suppliers and payments received from customers do not coincide in time, a certain amount of cash is always required to allow for the time delay between expenses and income for products or services.

Elementary ratio analysis was seen in the vertical analysis of SoftCo's comparative income statement. Indicators such as cost of sales as a percentage of sales can show how efficient different aspects of the business are.

3.17.1   Working Capital and Current Ratio

The working capital is the amount of current assets less current liabilities. A company needs working capital to meet current debts, carry inventory, pay employees, and take advantage of cash discounts with suppliers. A company that runs low on working capital is in danger of failing to continue operations. The amount of working capital required depends on the current liabilities (Fig. 3.10). The current ratio relates current assets to current liabilities as follows:

$$\text{Current ratio} = \frac{\text{current assets}}{\text{current liabilities}} \qquad (3.1)$$

Drawing on the data contained in the balance sheet (Fig. 3.11), SoftCo shows:

The high current ratio indicates that SoftCo is highly liquid. A current ratio that is too high (>10) may indicate that a company has invested too highly in current assets compared with its current obligations. SoftCo looks like a very good credit risk in the short-term, but the heavy investment in assets may not be the most efficient use of funds. A service company that grants little or no credit and carries little inventory other than office supplies, might function quite well with a current ratio of 1 to 1, while a high-end furniture store might require a higher ratio to be able to accumulate a large enough inventory to stock a large showroom with a large variety of items for its customers.

3.17.2 Acid-Test Ratio

Some assets can be liquidated faster than others. Cash, cash equivalents, and short-term investments are more liquid than inventory and merchandise. The acid-test ratio or quick ratio is a more rigorous way of testing whether a company can pay its short-term debts. Quick assets are defined as cash, cash equivalents, temporary investments, accounts receivable, and notes receivable. It is important to note that inventory is not included as a quick asset.

$$\text{Acid-test ratio} = \frac{\text{quick assets}}{\text{current liabilities}} \qquad (3.2)$$

Using data from SoftCo's balance sheet (Fig. 3.11), the acid-test ratio calculation is shown in Fig. 3.12:

(in thousands)	2005	2004
Current assets	$161,657	$113,203
Current liabilities	$25,084	$20,331
Working capital	$136,573	$92,872
Current ratio		
$161,657/$25,084 = 6.44 to 1		
$113,203/$20,331 =		5.57 to 1

Figure 3.10   Working Capital and Current Ratio for SoftCo

SoftCo Inc.
Comparative balance sheet
For years ending Jul 31, 2004 and 2005

Assets	2005	2004
Current assets:		
Cash and short-term investments	90,136	64,731
Account receivable		
Trade	52,568	37,411
Other	2,438	2,189
Inventory	15,069	8,031
Prepaid expenses	1,447	840
Total current assets	$161,657	$113,203
Capital assets	41,569	29,964
Total assets	$203,226	$143,167
Liabilities		
Current liabilities		
Accounts payable	8,554	5,419
Accrued liabilities	11,287	6,870
Income taxes payable	5,244	8,042
Total current liabilities	$25,084	$20,331
Deferred income taxes	2,446	2,491
Total liabilities	$27,530	$22,821
Shareholder's equity		
Common shares	101,711	73,776
Contribution surplus	376	377
Foreign currency adjustment	0	387
Retained earnings	73,609	45,806
Total shareholder's equity	$175,696	$120,346
Total liabilities and shareholder's equity	$203,226	$143,167

Figure 3.11   Comparative Balance Sheet

Once again, the test shows that SoftCo is extremely liquid, but a more complete analysis is required to examine how fast the company turns its assets into cash. We need to look at how fast the company can convert inventory into sales and then into cash.

3.17.3 Merchandise Inventory Turnover

This ratio indicates how long a company hangs onto inventory before turning it into a sale to a customer, and shows how efficiently the manufacturing part of the business is working. Service companies will generally have little in the form of inventory, but the ratio still indicates the efficiency

(in thousands)	2005	2004
Cash and short-term investments	$90,136	$64,731
Accounts receivable, trade	$52,568	$37,411
Total quick assets	$142,704	$102,142
Current liabilities	$25,084	$20,331
Acid-test ratio		
$142,704/$25,804 = 5.68 to 1		
$102,142/$20,331 =		5.02 to 1

Figure 3.12   Acid-Test Ratio for SoftCo

of the business process. Capacity planning and market demand knowledge can be used to improve turnover. It is defined as:

$$\text{Inventory turnover} = \frac{\text{cost of goods sold}}{\text{average merchandise inventory}} \quad (3.3)$$

For SoftCo, we have used the average of the inventory at the end of the reporting period in the denominator as an approximation to the average level of inventory instead of averaging the individual days for each piece of inventory.

$$\frac{\$37,103}{\left(\dfrac{(\$13,417 + \$7,361)}{2}\right)} = 3.57 \text{ times per year} \quad (3.4)$$

When inventory is "turned over" into sales faster, this ratio will increase.

3.17.4 Days' Sales in Inventory

For most people, a more meaningful ratio than inventory turnover is calculated by taking the amount of dollars in merchandise inventory at the end of the year and dividing it by the amount of money used to purchase the raw goods in that year. This is the proportion of dollars that are "tied up" before becoming available for sale. Multiplying this ratio by 365 gives the average number of days that cash has been paid out to suppliers, but not yet converted into a saleable article.

$$\text{Days' sales in inventory} = \frac{\text{ending inventory}}{\text{cost of goods sold}} \times 365 \quad (3.5)$$

Using the net sales for 2005 from the income statement and inventory from the balance sheet we get:

$$\frac{\$15,069}{\$37,103} \times 365 = 148 \text{ days} \qquad (3.6)$$

The data from 2004, however, shows that it used to take about 115 days to convert inventory to saleable merchandise.

$$\frac{\$8,031}{\$25,419} \times 365 = 115 \text{ days} \qquad (3.7)$$

The company has been experiencing very rapid growth and these calculations are quite coarse, so an investigation into the manufacturing value chain would best be conducted on a transactional level.

3.17.5 Accounts Receivable (A/R) Turnover

The next step following sale of merchandise to customers is the collection of cash from the sale. The A/R turnover ratio reflects the efficiency of this process. In a service company with little inventory, but with a large amount capital outlay for salaries, the collection of cash from customers is critical to the success of the business. The business processes that have a direct influence are terms and conditions for collection on contracts, correct billing information, accurate tracking of vendors' purchase orders, schedule of payments for partial payments, and so on. The turnover is defined as:

$$\text{Account receivable turnover} = \frac{\text{net sales}}{\text{average accounts receivable}} \qquad (3.8)$$

Some people use only credit sales in the numerator, and if the average accounts receivable by month are not available, then the average amount of receivables at the end of the reporting period is used in the denominator. SoftCo's turnover is:

$$\frac{\$167,907}{\left(\frac{\$52,568 + \$37,411}{2}\right)} = 3.73 \text{ times per year} \qquad (3.9)$$

3.17.6 Days' Sales Uncollected

By analogy to days' sales in inventory, by taking the ending accounts receivable from the balance sheet and dividing it by the year's net sales from the income statement gives the proportion of sales that has not yet been collected from customers. Multiplying this ratio by 365 gives the average number of days that a dollar from sales has not been converted into cash.

$$\text{Days' sales uncollected} = \frac{\text{accounts receivable}}{\text{net sales}} \times 365 \qquad (3.10)$$

From the 2005 financial statements for SoftCo:

$$\frac{\$52,568}{\$167,907} \times 365 = 114 \text{ days} \qquad (3.11)$$

This has improved slightly since 2004 (126 days).

3.17.7 Total Asset Turnover

The four previous ratios discussed in this section address components of the assets of the company, receivables, and inventory. If either of these two assets exists for longer than the reporting period of the company, they must be declared as capital assets. The last ratio we present includes all assets of the company regardless of the category.

$$\text{Total asset turnover} = \frac{\text{net sales (or revenue)}}{\text{average total assets}} \qquad (3.12)$$

For SoftCo 2005, the ratio is:

$$\frac{\$167,907}{\left(\dfrac{(\$203,226 + \$143,167)}{2}\right)} = 0.969 \text{ times per year} \qquad (3.13)$$

We follow the usual practice of averaging the total assets at the beginning and end of the reporting period.

3.18 Profitability Analyses

Analysts and stakeholders are especially interested in the ability of a company to turn its assets efficiently into profits. *Profitability* is the ability of a company to generate an adequate return on invested capital. When we worked with a GE acquisition (See Section 3.9, "Reporting the Benefits—The Corporate Dashboard") we found that the previous management had mistaken a profitability problem for a capacity and cash flow problem. Comparisons of ratios should be done using some heuristics (profit should be positive) and benchmarking against similar industries. While a similar and high profit is seen with specialty goods and services, low profits are expected with commodity type industries (see Fig. 2.5).

3.18.1 Profit Margin

Profit margin and total asset turnover are the two components of overall operating efficiency. We have discussed the latter and will now discuss the former. The profit margin gives an indication of how well the company can earn net income from sales. It shows how sensitive net income is to changes in revenue or costs.

$$\text{Profit margin} = \frac{\text{net income}}{\text{net sales (or revenue)}} \times 100\% \qquad (3.14)$$

The income statement will itemize a large number of components that bring about an increase in net income. An increase in sales productivity will increase the numerator more than the denominator for a net increase in profit margin. Information from the income statement (Fig. 3.8 or Fig. 3.9) shows that for SoftCo the profit margins are:

$$\frac{\$32,843}{\$167,907} \times 100\% = 19.6\% \, (2005) \qquad (3.15)$$

$$\frac{\$21,150}{\$108,333} \times 100\% = 19.5\% \, (2004) \qquad (3.16)$$

The profit margin looks healthy, certainly greater than the return from fixed rate government bonds, and steady from 2004 to 2005 despite a 55 percent increase in sales.

3.18.2 Operating Margin

Some analysts prefer to calculate profit in a different manner to focus more on the day-to-day operations of the company, independent of other factors such as interest expenses, interest income, one-time gains or losses, taxes, research and development, amortization, and depreciation. This ratio is calculated by taking the operating income, also called *income before interest and taxes* (IBIT), divided by net sales or revenue. It is used to evaluate the performance of companies with large amounts of debt and interest expenses.

$$\text{Operating margin} = \frac{\text{operating revenue} - \text{operating expenses}}{\text{net revenue}} \times 100\%$$

$$(3.17)$$

3.18.3 Return on Total Assets

The profit margin and total asset turnover can be combined to form a single summary ratio: the return on total assets.

$$\text{Return on total assets} = \frac{\text{net income}}{\text{average total assets}} \times 100\% \qquad (3.18)$$

SoftCo's results for 2005 are:

$$\frac{\$32{,}843}{\left(\dfrac{(\$203{,}226 + \$143{,}167)}{2} \right)} \times 100\% = 19.0\% \qquad (3.19)$$

Let us complete this section by showing the relationship between profit margin, total asset turnover and return on total assets. We start with Equation 3.18,

$$\text{Return on total assets} = \frac{\text{net income}}{\text{average total assets}} \times 100\% \qquad (3.20)$$

Multiply by unity:

$$\text{Return on total assets} = \frac{\text{net income}}{\text{average total assets}} \times \frac{\text{net revenue}}{\text{net revenue}} \times 100\% \qquad (3.21)$$

Rearrange:

$$\text{Return on total assets} = \frac{\text{net income}}{\text{net revenue}} \times 100\% \times \frac{\text{net revenue}}{\text{average total assets}} \qquad (3.22)$$

Simplify:

$$\text{Return on total assets} = \text{profit margin} \times \text{total asset turnover} \qquad (3.23)$$

Using Equation 3.23 and the results from the calculations of profit margin (Equation 3.14) and total asset turnover (Equation 3.13) for SoftCo for 2005:

$$\text{Return on total assets} = 19.6\% \times 0.969 = 19.0\% \qquad (3.24)$$

The same result was found at the beginning of this section using Equation 3.18.

3.19 Financial Benefits "Buckets"

As you complete the Define phase of your project, you should have gathered either the baseline financial benchmarks, or data required to do so. The baseline should be a period of about 12 months of financial performance for the business process you are attempting to improve. This must be the actual cost as could be found by a financial audit of the process.

These data must be tracked in some kind of system where benefits will be updated to the corporate dashboard automatically as your project proceeds

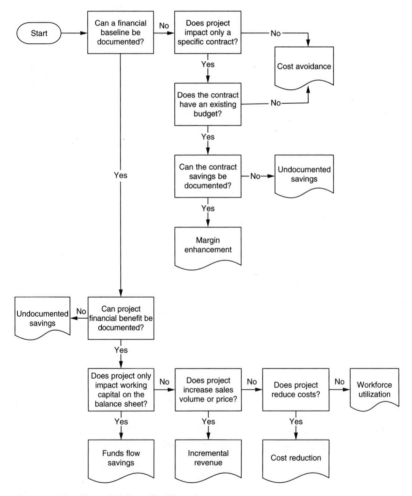

Figure 3.13   Financial Benefits Flowchart

towards completion. It is critical that the CFO and executive have defined the decision tree for "bucketing" financial benefits. Figure 3.13 outlines a suggested flowchart for classifying financial benefits into one of:

- Cash flow savings–frees up cash more quickly to be reinvested in the business. Lean Six Sigma, with the emphasis on speeding up processes, will have a major impact on cash flow savings.
- Incremental revenue–a project that permanently increases the sales revenue or volume. When processes become lean, the process capability frequently increases owing to less time spend on rework.
- Workforce utilization–the project improves the productivity of the existing workforce, usually by decreasing rework and unnecessary expedition of orders.
- Margin enhancement–usually only a single, large new contract for a customer, where baseline costs are not available, but present as a budget. The benefits are nonrecurring, because the next invocation of the process or contract will now follow the new business process.
- Cost avoidance–the new ability of rapid execution and delivery of defect-free services will now decrease the company's exposure to liquidated damages as a result of failing to meet contract terms for delivery, and so on.
- Undocumented savings–the savings may be estimated owing to low volume of the new services. The benefits may be estimated and entered in a tracking system and audited after a year. After the audit, the benefits may be reclassified into one of the other categories.

3.20 The Recognize Checklist

We have gone over a variety of subjects in this chapter to help the executive give quantifiable strategic direction to the MBBs, and for the MBBs to have a clear communication structure for reporting results up to the corporate dashboard for reporting lean Six Sigma benefits to external stakeholders.

The purpose of the Recognize phase is to identify the business needs and to articulate the problems in the business. There are three main stages in this phase.

3.20.1 Identify the Systematic Problems or Significant Gaps in Your Business

The source of ideas for changing your business can come from dissatisfied customers, systemic problems that have failed all previous attempts at solution, a changing marketplace due to changes in technology, customer expectations, the entry of new competitors, or a "burning platform" issue that is forcing the business leaders to make a difficult decision. You should be able to explain the real problems of the business in terms that everyone in the organization will understand.

At the beginning of this step:

- What are the problems in the business?
- What are you in business for?
- Why have previous business improvements or transformations failed to deliver the results you now require?
- Does everyone involved in the lean Six Sigma effort understand the business need?

At the end of this step:

- You have an honest appraisal of the business in the eyes of the stakeholders.
- Business leaders have acknowledged that there are problems.
- You have a clear description of the business need for change.

Points to remember:

- There are no sacred cows. Successful business transformations can involve letting go of entire business units that may have made money and good sense only a few years ago.
- Question all assumptions–be brutally honest.
- People will be very reluctant to reveal anything they feel may be used against them–identify and manage resistance.

3.20.2 Articulate the Business Strategy

Define the quantifiable deliverables carefully and completely to guide and measure project success and alignment. The executive management team and the board of directors should be meeting periodically to identify and commit to a set of objectives for the coming business year at the very least. Annual reports are a good source of ideas in general, but you will have to drill down to a greater depth of detail to turn strategy into execution. In general, a company should have revenue growth from year to year and be able to consistently deliver a better return on invested capital than rival industries. The senior executive should have a detailed execution plan.

At the beginning of this step:

- What do the directors want the company to deliver?
- What strategy should you follow to solve this problem?
- What are the details underneath the business problem?

At the end of this step:

- Does everyone inside and outside of the business support your lean Six Sigma program or project?
- Will everyone understand that the Six Sigma methodology has been proven thousands of times in hundreds of companies?
- Will the shareholders recognize your strategy?
- Is your strategy articulated clearly and specifically?

Points to remember:

- Stay clear of "just do it" solutions and strategies, they have probably failed in the past. If it was easy to solve the problem, it would have been fixed ages ago.
- *Pet projects* are usually based on strategies that have failed in the past. If they come up, go back and focus on the business needs.
- *Trial projects* and off center discussion will result in lukewarm results for your entire program.

3.20.3 Manage the Effort

Have the stakeholders agreed to your lean Six Sigma program plan strategy. Imagine standing in front of the shareholders at the annual meeting. In the first two stages of the Recognize phase you have outlined the burning issues of the business and have described what results you intend to deliver. The operational leaders of the business will now want to know how you are going to achieve these objectives.

If the implementation of lean Six Sigma or other process improvement methodology is large, then this step would explain how you are going to implement an entire program. This means determining the elements of quality leaders, MBBs and BBs, training, succession planning, reporting structure and infrastructure, project reviews, metric targets, and an extended rollout plan for transforming the company. If your implementation of lean Six Sigma is small, then this step will be a much smaller version of project management that includes limited parts of this large process.

At the beginning of this step:

- Are the targets and deliverables based in a realistic appraisal of the business?
- Are the senior management or project champions committed to the lean Six Sigma program?

At the end of this step:

- You will have a clear program plan including
- A corporate lean Six Sigma communication plan
- A career track for BBs into and out of the program
- Incentive and compensation program with bonuses or stock options tied to lean Six Sigma benefits
- A series of tollgates and reviews for training and project reviews
- A list of deliverables–people to train, BB reporting structure
- Identified and secured the resources to achieve the project goals
- A mitigation plan that addresses negative consequences of the lean Six Sigma project plan

Points to remember:

- Program failure will result if you neglect these first steps.
- If you encounter resistance later in your project, return to this Recognize phase.
- Find ways to show proof of concept, cite other Six Sigma success stories within or outside the company.
- Concentrate on a clear understanding of the problems and their impact on the business–steer clear of solutions.

4

Define

4.1 Variance to Customer Want (VTW)

In the beginning of Chap. 3, we presented the summary of GE's initiative for managing span on customer deliverables. The most important point was that customers did not want things quickly, but wanted them when they expected them.

We can illustrate the relationship between cycle time and customer expectations with a service example. The Canadian passport office in Calgary manages a large number of applications each day where the waiting time can be as long as one hour. An automated system generates a numbered ticket for applicants as they arrive. As passport officers become available, the ID number of the next applicant in the queue is displayed on a display board. If you do not respond within 10 to 15 seconds of your number being displayed, you lose your place in the lineup and must return to the dispenser and get a new ticket (Fig. 4.1).

The most important features of the system, from the viewpoint of the customer, is that the printed ticket shows where you are with respect to other people waiting for service and gives you the estimated time when you will be served. An accurate estimate allows you to plan to go for a cup of coffee, feed the parking meter, or conduct other business in the area. In the three or four times that the author used the passport office, the estimates were remarkably close.

A mature system of managing customers waiting for service requires knowledge of your service time for a variety of service offerings, the time dependence of customer demand, and management of your resources to meet customer demands. The interrelationship between the different components of measuring *variance to customer want* (VTW) are shown in Fig. 4.2.

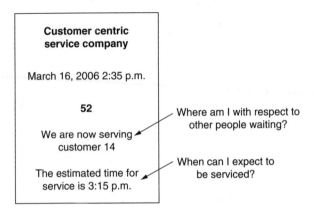

Figure 4.1 Take a Ticket for Better Service

The "span" shown in Fig. 4.2 is a measure of the width of the VTW that does not assume any particular type of statistical distribution. Very few sets of data measuring VTW follow any regular distributions. The definition of span used at GE was the width that included 90 percent of the distribution. It was also called the *P5-P95* to more precisely specify the width of the distribution between the 5th and 95th percentiles. The span also has much more meaning to customers than *defects per million opportunity* (DPMO) or a Sigma level. We will talk much more about the measurement of span and VTW in the Chap. 6.

An example of the distribution of VTW is shown in flight arrival times at the Calgary International Airport (Fig. 4.3). The figure summarizes the flight arrivals for 128 flights on a single day. The horizontal axis is minutes early

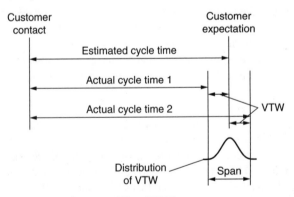

Figure 4.2 Variance to Customer Want (VTW)

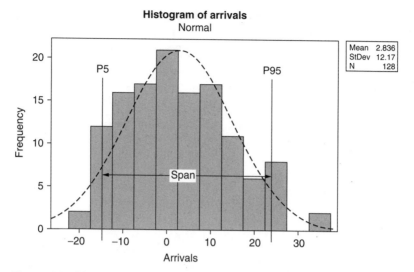

Figure 4.3   Distribution of Variance to Customer Want for Flight Arrival Times (*YYC, September 19*)

to the left and minutes late to the right. Simply reporting that the flights were late, an average of 2.8 minutes is not what the customers are interested in. Reporting the results as span would be stated as, "90 percent of our flights arrived between 14 minutes early to 24 minutes late."

The span measurement is much more meaningful than the often reported figure of percentage on time for airplane departures and arrivals. If a passenger was planning on transferring to another flight, they would be interested in the variance of the announced arrival times to make sure they could make a particular connection. People coming to the airport to pick up incoming passengers would be interested in estimating how much money they should put in their parking meter.

A valid question would be whether the span is caused by variation in estimated versus flight time as suggested in Fig. 4.2, or whether there is variation in departure times.

4.2 Defects—Six Sigma

The philosophy of Six Sigma from the original days at Motorola is that the output of every process is a function of the inputs to that process. In other words, what you do everyday in the business directly determines what the customers see. A major focus of a Six Sigma project is to establish the mathematical relationship between the defects produced by a business process and the causes of those defects. This is expressed as

$$Y = f(x_1, x_2, x_3, \ldots, x_n) \tag{4.1}$$

where, Y is the output to the customer and the Xs are the business processes. In a traditional Six Sigma project, a key component of the project charter is the operational definition of a defect. It is essential that this does not change throughout the project.

In the Measure phase, the number of defects generated by the current process will be measured by counting them (discrete data) or making a physical measurement of some kind (continuous data), examining the distribution of the data and calculating the process capability using the customer's *upper specification limit* (USL) and *lower specification limit* (LSL). The defect level is expressed as the DPMO and as the corresponding process capability (Z units or Sigma level).

In this context, a defect always refers to the resultant output of the business process that is unacceptable to the customer. The enumeration of the defect level involves a measure of capability of the business process to produce a product or service within the LSL and USL as set out by the customer.

The calculation of the traditional Six Sigma defect level (Fig. 4.4), involves collecting data, calculating the mean and standard deviation, then using the USL and LSL along with the equation of the normal distribution (Fig. 4.5) to calculate the proportion of the curve that is below the LSL, the proportion of the curve that is above the USL, and expressing the total proportion as DPMO and a Z-value.

The Z-value is the number of standard deviations away from the mean, enumerated by the DPMO. There is also a corresponding calculation for discrete defect counts. These calculations mean very little to customers and are fraught with assumptions that are frequently violated, making the calculation suspect.

The principal use for the process capability calculation is to quantify the theoretical defect reduction after the Six Sigma project has been completed. Owing to the nature of the calculation it is possible to reduce the DPMO while having no perceivable effect on the real defect level experienced by the customer.

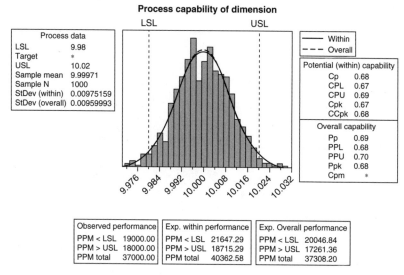

Process capability of dimension

Process data	
LSL	9.98
Target	*
USL	10.02
Sample mean	9.99971
Sample N	1000
StDev (within)	0.00975159
StDev (overall)	0.00959993

Within ——
Overall - - -

Potential (within) capability	
Cp	0.68
CPL	0.67
CPU	0.69
Cpk	0.67
CCpk	0.68

Overall capability	
Pp	0.69
PPL	0.68
PPU	0.70
Ppk	0.68
Cpm	*

Observed performance		Exp. within performance		Exp. Overall performance	
PPM < LSL	19000.00	PPM < LSL	21647.29	PPM < LSL	20046.84
PPM > USL	18000.00	PPM > USL	18715.29	PPM > USL	17261.36
PPM total	37000.00	PPM total	40362.58	PPM total	37308.20

Figure 4.4   Traditional Six Sigma Defect Definition

A set of data has been generated for a fictional technical service call center where technical queries are promised to customers at a particular time. The time the customer received the response is tracked and summarized. The set of data could represent the generation of quotes or keeping appointments in a medical clinic. The Six Sigma metrics have been calculated before and after a process change using a LSL of minus 30 minutes and an USL of plus 30 minutes. The results look encouraging, the DPMO metric has decreased from 260,000 to 187,000 (Fig. 4.6 and Fig. 4.7).

We have also calculated the span on VTW and found it has not changed. Fig. 4.8 shows that the *P5-P95* span has remained at 87 minutes after the process change.

Figure 4.5   The Equation of the Normal or Gaussian Distribution (*Zehn Deutsche Mark note*)

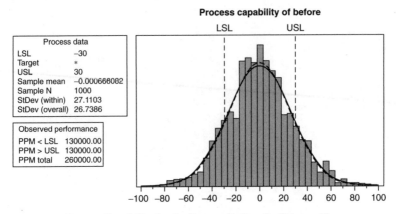

Figure 4.6   Process Capability for the Process Before the Process Change

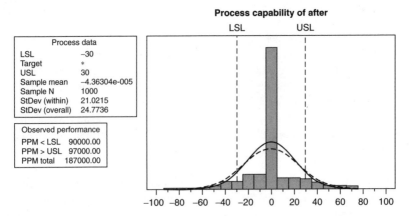

Figure 4.7   Process Capability for the Process After the Process Change

	Mean	**StdDev**	**DPMO**	**P5**	**P95**	**Span**
Before	0	26.7	260.000	−42.4	44.6	87.1
After	0	24.8	187.000	−41.4	45.7	87.1

Figure 4.8   Before and After Metrics for Order Processing

It is common that an organization will adopt Six Sigma, execute multiple projects, and report financial benefits to the shareholders while the customers do not feel the effects of the process changes. In the above example, the traditional Six Sigma metrics indicated to the improvement team that the new process had a 28 percent decrease in defects. The customers, on the other hand, would insist that 90 percent of the service is still anywhere from 42 minutes early to 45 minutes late. The customers are correct.

4.3 Defects—Lean Six Sigma

The performance of a process in lean Six Sigma is quantified by measuring the difference between the actual and expected elapsed times for the customer to receive a defect-free service or product. There are many advantages of using span as a metric:

- It is more relevant to making decisions about the process.
- It has meaning to the workers and customers.
- It is defined quite specifically with respect to the width of the process surrounding the customer expectation.

Defects are the reasons why a process has a wide span. In this sense the defects are the faults that cause problems in providing the expected service to the customers. It is the internal factors (Xs) that cause a large span on VTW (Y).

4.4 The Causes of Large Span

A business is a complex system and is not built up by such a simple, linear set of subprocesses to effect the outcome as was suggested in Section 4.2. We have already seen some of the multiple layers involved when we considered the Strategic Planning QFD in Chap. 3. When business processes are examined, there are multiple layers of relationships (Fig. 4.9).

Customer satisfaction involves the difference between the actual and expected elapsed times for the customer to receive a defect-free service or product (Fig. 4.2 and Fig. 4.9).

An essential part of lean Six Sigma is to how customer demand determines the business process capability. A misunderstanding of either customer expectations or process capability will lead to a wide VTW. There is potential for management to impose short-term solutions that solve only one half of the problem. Real businesses process frequently show problems with customer expectations and process capability.

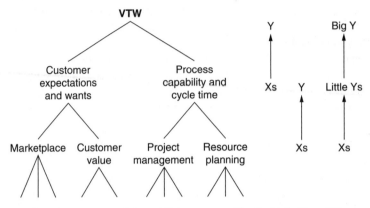

Figure 4.9   Business Processes and Subprocesses, Big *Y*s, Little *Y*s and *X*s

4.4.1 Variation in Cycle Time and Process Capability—the Little Y

Consider a service where the customer expectations are that the service will take 14 days. Customer orders come in on a random basis. The internal processes of the business are such that execution time is not always predictable and the result is that order processing and delivery, while taking about 14 days, sometimes takes longer or shorter (Fig. 4.10).

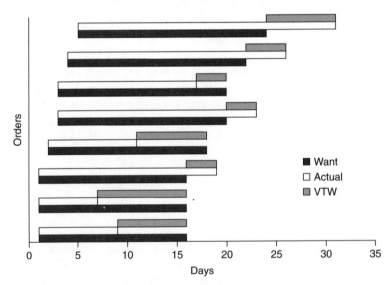

Figure 4.10   The Variation in Actual Execution Time Results in a Span on VTW. Each Job is Expected to Take 14 Days, but Actual Times Differ Between Orders

The *P5-P95* span on VTW is 14.2 days in this case. A lean Six Sigma project focused on reducing the cycle would have no effect on this metric. The median execution time is 14 days, but there is a problem with the variation in the execution time for orders being processed simultaneously. The cause of variation in VTW is the inability of the business to manage multiple orders, each with its own internal deadline.

4.4.2   Variation in Customer Want—the Other Little Y

Consider the same service organization where a lean Six Sigma project has resulted in an actual execution time of 14 days with no variation. Customer expectations, however, are not constant in this case. The cause of the variation in VTW is the inability of the business processes to respond to customer demand (Fig. 4.11).

After you have completed the Improve phase of your project, you will have to gather the same *Y* data, but now on the improved process. You must be able to prove that you have made a real improvement in the process by showing a decrease in the average cycle time and variation.

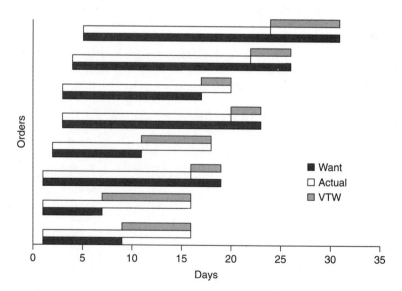

Figure 4.11   The Variation in Customer Expectations Results in a Large VTW. Each Order Takes 14 Days to Process, But Customer Expectations Differ for Each Order

In lean Six Sigma one metric we will be measuring is the total cycle time for the customer to complete a transaction of value with your company. This will help determine how much of a resource management problem (Fig. 4.10) versus a reaction to customer expectations problem (Fig. 4.11) you have. Defects occur, but they are one of the causes of delay and rework, or contribute to the cost of the product, thereby decreasing its value.

4.5 Mortgage Processing Time—Win/Win

When you are in the middle of data analysis you will be spending a considerable amount of time examining the dozens of reasons why the cycle time is as unpredictable as it is. These are the input factors or Xs of your process. Consider a mortgage application form being filled out at a branch bank before being forwarded to the central underwriting facility. It is important to see that the defect reduction in each of the Xs caused a reduction in the median cycle time. The span in cycle time was also reduced (Fig. 4.12).

Once the cycle time has been reduced, then the impact on the customers may be assessed by customer survey. In Fig. 4.12, the number of customers who left before the application was complete has decreased, presumably because the process has been sped up and does not require them to have as many extra meetings to correct erroneous or incomplete information.

The other consequence of faster cycle time for the mortgage application process is that the bank is now receiving mortgage payments more quickly from the customers. Oddly enough, this is not as bad for the customers as

Project: Mortage application processing

		Before	After
Y (Process output)	Cycle time	median - 25.4 min span - 23.3 min	median - 12.3 min span - 13.5 min
Xs (Factors or process inputs)	Missing or incorrect dates	12.5%	2.5%
	Missing signatures	8.5%	3.5%
	Wrong routing	1.5%	1.5%
	Missing birthdate	3.4%	2.5%
	Customer left before completing application	5.0%	0.5%
	Interest rate change	15.0%	1.0%

Figure 4.12 Mortgage Processing Metrics—Before and After

you might think. The customers have planned for, and expect the payments to occur. If the application is delayed too much they may not make the sale of the house, or have to pay a late penalty. Either is undesirable for the customer.

There are three distinct groups of data that will be gathered for your project.

1. *The project Y*: This is the measure of the output of the process and is sometimes called *critical to quality* (CTQ). In transactional lean projects it will be the span on VTW for cycle time of the service or product from the viewpoint of the customer. There should be only one Y per project. The change in this CTQ will be documented at the conclusion of the project.
2. *The potential Xs*: These are the internal factors that may influence the output Y. Brainstorming is a good technique for generating what can be a very large list. Do not concentrate on only the data that is currently tracked, you may have to devise ways of tracking new data.
3. *The financial impact*: When processes proceed more quickly, this decrease in time to market may result in increased sales or decreased abandonment rates. This must be documented somehow and included in your final financial impact conclusions. In a manufacturing lean project, it is easy to see that a decrease in physical inventory results in more cash available to the business. In a transactional lean project, the impact is less obvious. Variation in delivery in the transactional world may result in discounts to the customer if the deliveries are early, or liquidated damages against you if the deliveries are late.

4.6 Scoping the Business Process and Defining Team Roles

In the very early stages of the project, the business process should be sketched out at very low resolution; no more than about five steps to help orient the team. No detailed business process mapping should be done at this stage; it is merely to define for the team what is going to be the subject of the project. The details of issues that are part of the project are determined and assigned to team members. This meeting can go quickly and can give the entire team a good foundation in what is being done by whom (Fig. 4.13). Proceed by:

1. Sketch out the business process at the top of a whiteboard at low resolution. About four to five major steps are sufficient to start the discussion. Draw two large picture frames under the process map.
2. Use the simplified business process to brainstorm about potential issues to be addressed. Do not 'edit' or discuss the ideas at the moment.

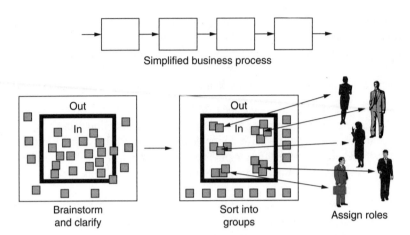

Figure 4.13   In-Frame/Out-of-Frame, Grouping and Assignment of Roles and Responsibilities

Have all the team members write their own issues down on individual Post-it notes, and place them on the whiteboard under the business process map. There may be well over fifty issues.

3. When the issues have been written down, take them one at a time and ask the team if any clarification is necessary. Do not proceed until all points are clear.

4. Review the issues one at a time and sort them into "In frame" and "Out of frame" regions of the first picture frame. You may have to alter the scope of the business process as you get a feel for the size of the project.

5. Sort the "In frame" issues into similar groupings. More than one person may have come up with a network security issue, for example.

6. Ask the team to look at the sorted categories and have everyone have a second round at defining issues. Some of the ideas from other team members may spark new ideas in the other members.

7. Get consensus that the issues are clear and complete, and that the exercise has clearly defined what is part of the project, and what is not.

8. The clusters of issues can be assigned to the team members according to their greatest strengths. Clusters of issues that are not well represented by the existing team members can be used to describe additional team members.

9. Record the project scope, roles, and responsibilities. Use this to define the communication and execution plans for your lean Six Sigma project.

If the project seems too large, trim the scope a bit to get something that can potentially be completed in about 3 months, or use the clustered issues to define a multigenerational project plan.

4.7 Resistance

You may think that a section on dealing with resistance has no place in a statistically based book, but years of experience has shown the author that all the best technical information in the world does not automatically and irrevocably lead to the perfect solution.

One of our best lean Six Sigma projects started at a small service shop with three full time employees and a part-time bookkeeper. We concentrated on defining and implementing accurate and timely communication between the workers and customers. At each stage during the Improve phase, we would ask the workers if the new communication processes would make their work easier and solicited their input for modifications. When the new processes were adapted, productivity, collections, and customer satisfaction increased while rework and billing inaccuracy decreased.

The project had very little in the form of hard data and statistical analysis, but was happily accepted by the process owners. Other service shops started to share the results of the project "over the grapevine" and spread the practices. The project had a huge impact on the customers and the workplace. The project eventually became the backbone of a worldwide implementation for work scope definition and reporting. This can be summarized as:

$$\text{Effectiveness of solution} = \text{quality of solution} \times \text{acceptance} \quad (4.2)$$

When you are steering a project team, you will encounter resistance at all stages. You will probably not get the same type of resistance from the same people as you progress through your project, so you must continue to monitor and manage it. The objections you will hear will change as more information becomes available.

There are a number of broad categories for the reasons why people may be motivated to maintain the status quo (Fig. 4.14). Your team may have members with different personality types who are best suited to discussing the project with the people resisting. Personality type profiling, such as those based on the Jung-Myers-Briggs topology, can be helpful in determining which team member is best suited for communicating with different stakeholders. It would be a poor choice to have a highly technical member talk with someone worried about loss of union jobs. For the most part, power-based pressure to comply with the project is a poor motivator and will eventually lead to poor results (Equation 4.2). We have often found true value in listening to and understanding the objections. The people voicing their

Resistance type	Reasons	Team type for communication
Cognitive	People truly believe, based on their own information and experience, that the diagnosis (why change?) is wrong and the proposed course of action (lean Six Sigma Project) is misguided	Technical
Ideological	People believe that the proposed change violates fundamental values that have made the organization what it is	Technical symbolic
Power driven	People perceive that for them, the proposed change will lead to a loss of power, autonomy and self-control, that it will lead to reduced status and autonomy	Political
Psychological	People have difficulties to learn, adopt or assimilate new concepts and new behaviors	Political symbolic
Fear	New situations and unfamiliar processes can lead to a general, unverbalized fear of changing the ways things are done. It will eventually lead to one of the above types if not addressed	Political symbolic technical

Figure 4.14   Resistance Types and Strategies

concerns have usually seen a number of past attempts at process improvement and are convinced that new initiatives are the management flavor of the month. Previous efforts may have failed owing to lack of input from the process owners. One of the more successful strategies we have used is to offer to include the objectors in the improvement team.

Resistance has been described as progressing through five or six stages. This is not to say that resistance will increase, but more that the objections you will encounter will evolve as you take down each barrier. This progression has been called variously as the levels of resistance and the levels of buy-in (Fig. 4.15).

Type	Action required
1. You do not understand my problem	Listen, do a thorough assessment of the present situation. Include finding out about solutions that have been tried before, but failed. Do not blame anyone for past failures, use the experience to learn
2. We do not agree on the direction of the solution	You understand the problem now, but the team has trouble with agreeing on the direction of the solution. Take care to exclude the solution in search of a problem. Previous attempts at solving the problem may have still have stakeholders convinced that their solution only failed due to lack of resources
3. We do not agree that your solution will have the impact you predict	Go back to the original description of the process with your cause and effect diagram showing the causes of defects. Assess if you have missed any possible causes of defects or if the data analysis is valid
4. Your solution is going to interfere with some existing initiatives	At the beginning of your Define stage, make sure you have listed these other inititives, cost cutting, big marketing compaigns, recent large growth. Check your in-frame-out-of-frame project definition to make sure the edges of your project have been maintained. This may be an opportunity for power based resistance types to try to claim improvements under the umbrella of another initiative. Some changes will have some negative impact on the business, try to quantify it and balance it against the benefits
5. There are some large and real problems with the proposed implementation	These objections could be related to budget, legal problems, union labor, financial risk, large equipment purchase, extensive retraining, and so on. You should be prepared to defend the financial benefits and customer impact for each improvement alternative. Management may have a number of other projects competing for limited resources
6. We have to do more analysis, wait and see if it improves by itself, and so on	This is not really a level of resistance, but more the remainder if the previous levels have not been addressed completely. There may be some remaining fear that shows up in an unpredictable manner. Powerful, but mostly silent stakeholders may still be a few levels back in resistance. Go back and make sure you have dealt with each level of resistance with each stakeholder group

Figure 4.15 The Six Levels of Buy-In

Project: S1245
Pricing accuracy

Stakeholder buy-in tracking

Team	
MBB	J.deV.
BB	S.L.M
Controller	A.K.M
Finance rep (A/R)	X.L.
Service rep	J.W.

		Buy-in stage				
		1	2	3	4	5
Stakeholder		You don't understand the problem	We don't agree on the direction of the solution	Your solution will not have the predicted impact	This will interfere with existing projects	There are problems with the implementation plan
V.P. sales	Completed	4/28/2005	5/12/2005	5/12/2005		
	Team member	MBB	Controller	BB		
	Action items	Invited to team meeting	To present alternate plans at team meeting	Email margin by product line control charts		
CFO	Completed	4/30/2005	5/2/2005	4/30/2005		
	Team member	Finance rep	Finance rep	MBB		
	Action items	Discuss customer satisfaction survey	No further action required	Invited to team meeting		
Service manager	Completed	4/30/2005	4/30/2005	4/30/2005		
	Team member	BB	Service rep	Service rep		
	Action items	Discuss customer satisfaction survey and past improvement projects	No further action required	No further action required		
Quality leader	Completed	4/28/2005	4/30/2005	4/30/2005	4/30/2005	
	Team member	MBB	MBB	MBB	MBB	
	Action items	Discuss customer satisfaction survey and past improvement projects	No further action reqiured	No further action required	Discuss project goals with project portfolio	

Figure 4.16   Managing the Five Levels of Buy-In for Different Stakeholders

During the early stages of negotiations of large sales contracts, we have seen the sales teams assemble a strategic plan based on the resistance levels with risk assessment and mediation plans for each step. When these are combined with different stakeholder groups composed of different resistance types, resistance plans can become quite complex, but quite successful (Fig. 4.16).

The buy-in plan is updated at each team meeting when resistance from the stakeholders can be evaluated by the team; the best person is assigned to that stakeholder and the remediation steps are taken.

4.8   Managing the Project Team—the GRPI Model

The stakeholder analysis is a good tool for managing resistance from outside your team. The *goals, roles and responsibilities, processes and procedures, and interpersonal relationships* (GRPI) tool helps manage a successful team over the length of the project. The project scoping exercise

(See Section 4.6) has helped you define the initial start of the project. The roles and responsibilities can continue as input in the GRPI model. The GRPI model may also be useful as a diagnostic tool when the team is not working well and you are not sure what is wrong.

The GRPI model periodically assesses the status of the project with respect to:

- *Goals*–The goal should be *specific, measurable, attainable, relevant and timely* (SMART) and included in the project charter. Each member of the team should understand and be able to explain the project's goal in their own terms. Are the goals of the team clear and accepted by all the members? If the GRPI model is being used periodically at team meetings, then the question is, "Are the items on the action item list clear and accepted by all team members?"
- *Roles*–There are many aspects of the business process that could make an impact on the goal. Team members should know what parts they are responsible for and which parts the other members are responsible for. The project scoping exercise should have identified any missing competencies and resources, but the project scope can change with time. When the GRPI model is being used periodically, then ask the team members if the items on the action item list have been assigned and accepted by the team.
- *Processes*–This does not refer to the business process the team is working to improve, but rather the business processes for managing the meetings, communication, conflict resolution, resource allocation, and decision making.
- *Interpersonal*–It may be necessary to itemize the rules of conduct for the team. Some teams outline the decision making process, then the project leader vetoes decisions by appealing to higher authorities. The team should be a safe place where communication is open, impartial, and respected. The team should be open and flexible in adapting the goals, processes, and roles with input from all its members. The interpersonal perspective is considered last because it tends to be supported by the other three portions.

We have seen some projects where a GRPI survey was individually completed by all the team members at the end of each team meeting. They would score the project on each of the four perspectives using a scale of 1 to 10. The summary scores were charted and circulated to the team as part of the communication plan.

Initially the goals, roles, and processes are the most important to define, but as the project progresses, the interpersonal state of the team can quickly become the greatest concern for continued success.

4.9 The Define Checklist

The executive has given you strategic guidance and performance goals and metrics for your lean Six Sigma project portfolio. The Define stage shifts your viewpoint to that of your customer. There are three main stages in this phase.

4.9.1 Change Your Viewpoint

You must capture the viewpoint of the customer in the purest possible form. This may be in the form of customer surveys, feedback forms, marketplace measures, or other feedback. Your customers should be able to recognize these measures and agree that they reflect their view. Success in your project will be determined by the customers' view of your project, not your manager's. Identifying the customer is not always straightforward in business-to-business transactions. Is it the CEO, the sourcing department, the engineers, or the receiving dock.?

At the beginning of this step:

- Who are your customers?
- How do you make an impact on their business processes?
- Do you understand these effects from the viewpoint of average and variance?

At the end of this step:

- Will your customers recognize your list of CTQs?
- Are you working on the most important customer problem first?
- What sources of information and data will you use?
- What do you presently do to track performance? Is this a cause of the problem? Will it have to be changed?
- Is there a good connection between the internal measures and the customers' views?
- Do you have a clear, quantifiable, operational defect definition?

Points to remember:

- Beware of conflicting customer views.
- Do not try to interpret or translate the customers' views.
- Accept what you are–the goal is to improve, not to cover up or gloss over.
- Your defect definition will not change during your entire project, unless the project changes.
- Validate your assumptions, with your customers.

4.9.2 Develop the Team Charter

It should be clear at the onset what the team members are going to be doing and how communication is going to work. The project is going to have a limited lifetime and team members will be engaged in other projects at the same time as this lean Six Sigma project. You may have to include team members' managers in the communication plan to deal with resistance. The *team charter* is a written document that defines the roles and responsibilities. At this point, there is only an estimate for the magnitude of the problem and the impact of the solution. During the Measure phase the data required to quantify the problem will be gathered.

This charter will continue to change as data analysis is conducted throughout the project. More team members may be required as the business process is dissected. The charter is part of your communication plan and should be shared and modified by the team.

At the beginning of this step you should have:

- A draft copy of the business process at "low resolution."
- Identified the primary job functions that are part of the business process.

At the end of this step you should:

- Have names associated with the job functions in your process map.
- Identified stages of resistance and developed a resistance plan.

Points to remember:

- You will continue to add to the charter when mapping, defining, and scoping the business process.
- People will be more committed to your project if they understand what is expected of them, and that they are part of the team effort.
- The project may change considerably if the project becomes a *Gage repeatability and reproducibility* (R&R) project during the Measure phase.
- You must have a financial team member if the project's financial impact is likely to be large or complex.

4.9.3 Scope the Business Process

When a dedicated and motivated team is assembled to tackle a complex business problem, the project can suffer from scope creep when others keep expanding the business process. Carefully determine what parts of the business process are in-scope and what parts are out-of-scope. You have to balance this tendency for projects to expand with the fact that business processes are infrequently well defined and contained to begin with. Three months is about the right length of time for a project—shorter times usually mean that the project is a trivial one or is not rigorous enough to sustain the improvement—longer ones make the project lose momentum with endless meetings and conflicting new tasks for your team.

If the project seems too large, it is. Ask the question, "If this is such an important problem for the business, why are we spending 6 to 9 months solving it?"

At the beginning of this step you should ask the questions:

- Have you determined who is on the team and are they all present at meetings?
- Are the process owners on the team? If not, why not? They know the process better than anyone else and will eventually own the new process.

At the end of this step you should:

- Have a clear understanding of individual issues by all team members.
- Have defined a project with an estimated time line of about 3 months.

Points to remember:

- Defining what parts of the project are out-of-scope may be as useful as defining what parts are in-scope.
- The list of issues may seem very large (>100), but it is essential that all ideas are considered.
- There is no problem solving to be done at this time. It is an ingrained habit of many managers to attempt to quickly solve problems as they appear or trivialize issues that make them uncomfortable.

5

Measure

5.1 Level of Detail

You have now defined your project. You have a commitment from management and a communication plan for the team. Part of the project charter is a measurement of the defect rate. In your previous traditional Six Sigma projects the defect definition might have been a discrete measurement, the number of applications with missing information or a continuous measurement, the difference between actual and estimated project costs. In your lean Six Sigma project you will be measuring the total cycle time for a transaction from the customer viewpoint.

This is one of the most difficult and critical steps for project success. You must step out of the usual definitions of business departments and functions. There will be problems with definitions and the scope of the transaction that may require you to change your project charter.

The definition and level of detail should make sense to your customer. Be particularly careful about consolidated measurements such as month-end averages reported at an upper business level. When people are fearful about the possibility of the data being used for individual performance evaluation, they can intentionally use averages to hide problems. If one-half of your deliveries in a month are one week early and the other half are one week late at the divisional-level, reporting the average will make it appear that all of the deliveries are on time at the corporate level. You must gather data at the level of the individual transactions and in a manner that would make sense to the customer. In Chap. 4, we showed that reporting weekly averages on variance to target can be misleading when your customers are measuring whether you are meeting delivery expectations on a daily basis.

The customer and the business commonly have different definitions of cycle time, thus driving a difference of perception about a process. This is usually

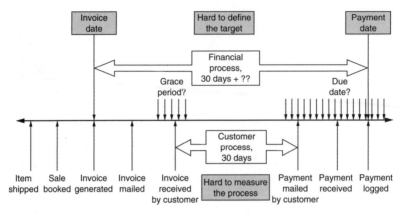

Figure 5.1   The Definition of "Accounts Due In 30 Days" Means Different Things to Different People

caused by management making business decisions based on easily accessible data rather than data that truly reflects the customers' viewpoint. A deceptively simple term is "accounts due in 30 days." Figure 5.1 shows an example drawn from a project on cycle time reduction of accounts receivable. Owing to the *silo mentality* in the business, the sales team would renegotiate payment terms for individual sales, but neglect to reliably pass the information onto the finance department. Internal delays in processing payments would also generate overdue notices for payments that had already been received but not logged against the account.

There is always a temptation to define cycle time in a narrow sense and argue that a more encompassing definition is outside the scope of the project charter. Figure 5.1 shows a number of processes that could easily be defined as out-of-scope: delivery by the postal service is clearly out-of-scope, but must be included in the definition of cycle time because of the customers' perception of "accounts due in 30 days."

Whether an income tax return is deemed late is determined by the postmark, regardless of when the specific government agency office concerned eventually received it. The receipt of a tax return by an agent of the government is recognized by both the customer and the business as a clearly defined point when the transaction has passed from the customer to the business.

5.2 Process Mapping

The graphical representation of a manufacturing process is a good communication and organizational tool when a process is linear, limited in physical location, and relatively specialized. Process mapping can become much more complex when the process is dynamically reconfigured for different product offerings or capacity demands. Like any mapping exercise, the result is a graphical representation of the process to be used for communication or diagnostics.

A rough process map showing the macroscopic steps of your business process was constructed during the Define phase. This process map was low in detail and diagnostic value, but it was a useful exercise that helped to define the project scope. More detailed process maps can be constructed to look at logistic restrictions, information flow, risk management, staffing and training requirements, costs, or inventory flow. Each process mapping exercise will have a specific purpose and emphasis.

Transactional processes are usually very complex and the team will tend to try to map only the majority of transactions to keep the process map from getting too large, treating a significant number of transactions as exceptions that should not be included in the analysis. Our experience is that these exceptions are quite common and are usually the source of the majority of the problems in the process. Making unrealistic promises to customers, assembling special orders, expedited shipping, and changing credit terms for an old customer are the sorts of things that create enormous impact on the rest of the "ordinary" transactions.

This may be the first time the process has been mapped in this much detail. The resultant map will most certainly look unlike any business perception that the process is simple and approximately linear. The team may feel some pressure to alter the map as a quick fix to avoid blaming a particular department. This is not a finger pointing exercise, but a diagnostic one to fix chronic problems in the design of the process, not in its execution. A common mode of resistance is a stakeholder who claims to be aware of a newly highlighted problem and claims to have already taken action to address it. This tends to undermine any conclusions the teams make using historical data. If the process has been changing through time, try to document the dates of the changes so you can flag historical data and treat them accordingly.

We ask that the proposed changes wait until the Improve phase where they can be considered along with all the other proposed changes.

The process map can go through so many cycles of mapping and validation that we usually construct it by starting with a large whiteboard and place sticky notes for process steps and use erasable markers for arrows—accept that the mapping will only eventually be documented using flowcharting software. Computer screens are too small for a team to work with and limit the visibility of the entire process.

An example of a process map for an order management system is shown in Fig. 5.2. It fits on a single page, is neatly organized in swim lanes, and emphasizes interdepartmental communication and resource allocation. This is an example of the process map for the ideal case without the errors and rework loops. These rework loops cause a majority of the work and create major cycle time delays.

Transactional processes have a slight disadvantage over manufacturing processes with respect to process mapping:

- The definition of an entity can be difficult because of consolidation of information as transactions move through the process.
- The flow of these entities through the process can be difficult to follow without reliable tracking systems.
- Many people may have different ways of performing with what appears to be the same action.

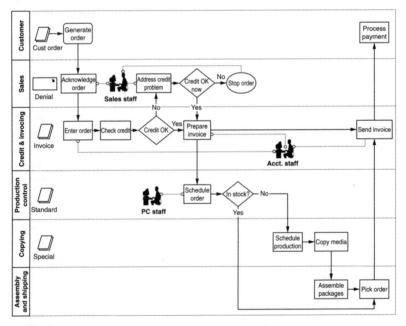

Figure 5.2 Process Map Emphasizing Interdepartmental Communication and Job Functions

- The entities do not visibly move through the business process the way items move through a physical manufacturing process.
- Changing a transactional process is easier than retooling a manufacturing process, so the processes tend to be altered frequently with a minimum of concern about the impact in the long-term or downstream in the process.
- Workers in the business may have multiple responsibilities in the process and see their tasks as a single job.

All business processes will change over time as incremental changes are made. The nature of transactional process will allow people to easily make many suboptimal changes without any visible effect. Transactional processes can become extremely complex before the process owners discover problems.

5.3 VA/NVA Process Mapping

The purpose of the process mapping exercise at this stage in your project is to document the process steps necessary for taking the transaction from the customer into the business processes and then bringing it back to the customer. When you are interviewing the subject matter experts, map the flow of the entity from the viewpoint of the customer. In other words do not ask them for their day-to-day job descriptions but ask questions in the form of, "What do you do with this order form?" Do not spend too much time defining departmental responsibilities or physical locations except where physical transportation of some kind is required.

This detailed process map can be analyzed in the sense that each process step can be classified as to whether the process step adds value to the service or product. Typical process steps that do not add value are inspection, transportation, delay, and storage. The *Value Added/Non-Value Added* (VA/NVA) process map can be used to classify and document whether a process is well designed to deliver service to the customer in an efficient manner.

The usual result of a VA/NVA process mapping analysis is a list of non-value added process steps that can be eliminated or automated to decrease the total cycle time. A necessary assumption is that each process step has a non-zero cycle time, so the elimination of steps must result in a decreased total cycle time.

This does not mean that every non-value adding process step should be eliminated. An example of a necessary non-value added process step is a regulatory requirement for reporting or approval. Even those, though, can be considered as a mitigation mechanism that adds value by reducing risk. Another example of a non-value added step would be a credit check as part of a loan application process. The customer would not usually agree that a credit check adds value to a loan, but eliminating the step would result in higher loan costs that would be ultimately passed on to the customer in the form of higher loan charges.

This mapping exercise is useful if this is early in the lean Six Sigma implementation at your company. The process map may be the first one that shows the flow of the service as it passes between the different organizational units of the company, and may highlight problems with roles and responsibilities for different parts of the process. There are frequently a great deal of "just do it" or "quick win" improvements that can be identified even without quantitative data. Other rapid opportunities for improvement can be made in the light of new technologies or capabilities.

The invisibility of transactional process contributes to the problem of hanging on to the old way of doing something after losing track of why the process was changed to begin with. Incremental changes can result in elaborate, but unnecessarily complex processes. One project had an approval process that was originally established as a communication process. We replaced the seven step approval process with a two-level approval process combined with a simple communication plan involving the balance of the stakeholders. The cycle time decreased substantially with no increase in risk.

Value added/non-value added process mapping does not always specifically highlight the causes of long cycle times caused by backlogs unless the material is transported to a different location or checked-in and -out of a storage state. Process maps appear much cleaner than reality because storage, rework, and delay times are hidden in the arrows connecting process steps.

5.4 Value Stream Mapping

Many of the problems associated with non-lean processes in manufacturing are caused by the management viewpoint that an idle machine or worker is a cost. This business centered perception leads to the belief that maximum resource utilization will reduce costs. This will, of course, result in manufacturing excess inventory that will ultimately result in long queues for intermediate steps and excess costs for carrying inventory. In transactional processes, the *work in process* (WIP) and cost of maintaining it are easily ignored because it does not build up physically. The result is long cycle times and a poor reaction time to changes in customer demand.

Value stream mapping for lean Six Sigma has a different emphasis than the simple retention or elimination of process steps seen in VA/NVA process mapping. If a process step takes very little time, then the impact of eliminating it is minimal. The main objective at this phase of your project is to measure the time required for all the process steps. Early process mapping exercises in transactional projects are often quite exploratory, requiring only estimates of time taken at each step for useful analysis, and data collection planning. Estimates of time data can be gathered by survey while the process is being mapped with the project subject matter experts. Use a scale for quantifying time such as the one shown in Fig. 5.3.

The initial evaluation of the process map can identify bottlenecks in the process and allow an estimate of the total execution time and cycle time. The assumption during this exercise is that all the process steps add value of some kind, and the delays between process steps are the largest source of non-value added time.

The value stream map summarizes the delay and execution times for an individually identifiable entity as it moves through the process and reflects the process at a particular time. All "normal" backlogs are reflected in the delay times. Inventory levels at each step can also be recorded and noted on the value stream map between process steps.

Typical delay and execution times have been added to a simple, four step value stream map for scanning and indexing medical claims documents (Fig. 5.4).

Note that these data are the elapsed times for individual, identifiable documents, and files. When the file is being reviewed, it does not take 3 days for the file to reach the reviewer, but it takes 3 days for the reviewer to finish the case load on her desk first, before considering the file of interest.

Rating	Time	Rating	Time
1	20 s	6	4 hr
2	1 min	7	8 hr
3	5 min	8	1 day
4	20 min	9	3 day
5	1 hr	10	7 day

Figure 5.3   An Empirical Time Scale Useful for Survey Responses During Process Mapping

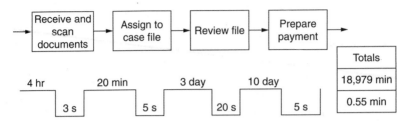

Figure 5.4   Value Stream Map for Medical Claims Processing

The times shown in Fig. 5.4 are not unusual for transactional processes. The totals for value added and non-value added time in this process show that vast majority of the time it takes to prepare a payment is delay time (99.997 percent). This delay could be caused by a large backlog, long changeover times, storage, rework, or transportation. The value stream map does not distinguish between different types of non-value added time and does not show rework. It is not common to collect data on the variation in times or the probability distributions they follow. Both of these characteristics cause problems in managing resources and capacity for downstream processes.

The process mapping becomes more difficult when entities are consolidated, split into different priority work streams, or when different parts of the process are executed with differing shifts or workdays. Other features that can be added to the value stream map are triangles summarizing average inventory levels in place of the single arrows. Under each process step can be a list of causes of problems for each process step, and information flow such as order signals. Where it is relevant, the delay time can be divided into changeover time, scheduled downtime with shift information, and batch size.

When lean manufacturing improvement techniques are applied to a process, it is usual that the backlog is significantly reduced. The execution times in Fig. 5.4 will not change remarkably, but the delay times will be reduced.

5.5 Flow Rate, Cycle Time, and Inventory—Little's Law

The emphasis in this book is to make the transition from traditional Six Sigma projects to lean Six Sigma projects. This means a shift from thinking about defect reduction to thinking in terms of increased process flow. The two concepts are combined in our lean Six Sigma defect definition of, "Reduce the time it takes to deliver a defect-free product to the customer." Individual transactions replace the concept of a manufactured unit.

There are three important metrics for flow measurements;

- The number of transactions that pass through a process step in a given unit of time.
- The time a transaction spends within the boundaries of the process step.
- The number of units within the boundaries of the process step at a given instant in time.

We will be spending more time on measuring the metrics for the entities entering and exiting the process step boundaries at the individual transaction level, but there are some basic observations we can make about the relationship between the averages of these quantities. The assumptions for the moment are that input queues are full, demand is fairly constant, and execution times are consistent.

Each process step either explicitly or implicitly contains a number of entities, either awaiting processing or being simultaneously processed. If a process step was a part of cooling down after heat treatment, then there would be multiple parts all being processed simultaneously. In transactional processes, it is common to partially process an entity and then place it in a WIP pile where it awaits missing information or approval before being processed further. This WIP is distinct from entities awaiting processing because the queue is full, but the data usually does not differentiate them (Fig. 5.5).

Look at the statistics gathered for the second step in the document scanning process for medical claims shown in Fig. 5.5. The process step "assign to case file" is divided into a queue of documents awaiting assignment and a group of documents that perhaps require clarification from a supervisor (WIP), or are treated as a batch at the end of an operator's shift. We do not have data about the number of documents delayed in the input queue or the number in WIP, but there are still some calculations we can perform to estimate the average number of documents awaiting completion.

Logically, if it takes an average of 1205 seconds (20 minutes + 5 seconds) for a document to pass though the process step, and it takes an average of 5 seconds to assign a document to a case file (rate = 0.2 documents/sec), then there should be an average of 241 documents "in process" at any one time.

$$1205 \text{ sec} \times 0.2 \text{ documents/sec} = 241 \text{ documents} \qquad (5.1)$$

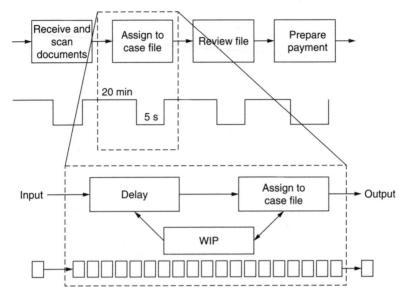

Figure 5.5 Process Mapping Metrics Do Not Usually Distinguish Between WIP or Delay

In the general case:

$$\text{Time} \times \text{rate} = \text{inventory} \qquad (5.2)$$

This relationship is known as *Little's Law* and is often used in value stream mapping. In principle, it is only necessary to collect two of the three quantities related to flow and use Little's Law to calculate the remaining one. The relationship is valid when the process is stable, but can give misleading results when supply, demand, or execution times are variable. In general, when inventory levels go up, cycle times increase: when the queue is longer, people have to wait longer for service.

5.6 Process Yield

The yield of a process is never going to be 100 percent at all times. This may be caused by transactions that get delayed, abandoned, lost, reworked, or reassigned to another process. The usual assumption is that the extended transaction is still taking place until a customer specifically cancels an order, hangs up the telephone, or walks out of your office. This is called different things by different industries—the abandoned shopping cart, hang-ups, or balking. The decision to abandon in the manufacturing arena is usually made by the manufacturer based on rework costs. The decision in the transactional arena is more commonly made by the customer based on excessive waiting time.

The calculation of process yield requires some clarification because it can reflect the initial process step, or the composite of process and rework (Fig. 5.6). The *classical yield* uses the ratio of output to input and ignores the details within the dotted rectangle shown in Fig. 5.6. Abandoned transactions are the only ones considered as defects, and all other transactions are counted as "good." This will overestimate the efficiency of the process because it does not distinguish between transactions that occur without error and transactions that only *eventually* occur without error. In a typical transactional process, the rework involved in correcting an error can be much more than the time of the error-free transaction. This recurring rework is referred to as the *hidden factory*. Classical yield has no discrimination power to identify problems with rework.

The *first time yield* calculation tracks only the proportion of transactions that proceed from input to output without error. The error-free transactions must be specifically tracked and some care should be taken to carefully track output from the *process* step in Fig. 5.6 to distinguish error-free transactions from transactions that have passed through the rework loop and are eventually correct.

The *rolled throughput yield* is a theoretical calculation based on the defect counts from all transactions. It assumes a Poisson distribution of discrete errors, and distinguishes between transactions having zero, one, two, or more defects, whereas first time yield only classifies transactions as defect-free or defective. The parameters from the calculation or rolled throughput yield can be used in process capability calculations and Monte Carlo simulations of the process.

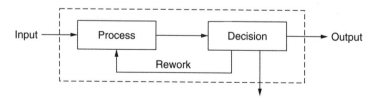

Figure 5.6   A Process with a Decision Resulting in Rework or Abandonment

The data from Fig. 5.7 can be used to illustrate the differences.

$$\text{Classical yield} = \frac{\text{output}}{\text{input}} = \frac{994}{1000} = 99.4\% \tag{5.3}$$

$$\text{First time yield} = \frac{\text{number with zero defects}}{\text{total number}}$$

$$= \frac{818}{1000} = 81.8\% \tag{5.4}$$

$$\text{Defects per unit (DPU)} = \frac{0(818) \times 1(164) \times 2(16) \times 3(2)}{1000}$$

$$= \frac{202}{1000} = 0.202\%$$

$$\text{Rolled throughput yield} = e^{-\text{DPU}} = e^{-(0.202)} = 81.7\% \tag{5.5}$$

Our experience is that data management systems track the transactional process steps when they are completed at an upper level; the input and output steps in Fig. 5.6 and Fig. 5.7. This will only allow the calculation of classical yield, and will grossly overestimate the true process yield.

Process improvements in traditional Six Sigma projects are assessed using metrics involving defects and yield. Lean Six Sigma, with its emphasis on cycle time, will use yield calculations, but only in the context that errors cause delays and result in inefficient processes.

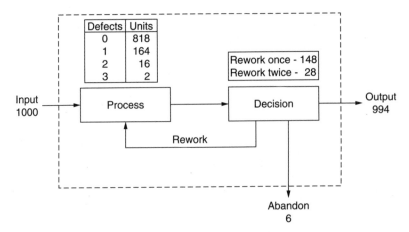

Figure 5.7   Data for Yield Calculations. Defect Data Does Not Include Reworked Transactions

5.7   Process Efficiency Mapping and the FMEA

A process is considered "lean" when the value-added time in the process is more than 25 percent of the total time of that process. Given that definition, the process shown in Fig. 5.4 is quite "fat." It is very common to have more than 90 percent of the processing time attributed to delays and rework, so keeping track of hand-offs between process steps is very important. We need to map the process in a manner that captures the execution time when the process proceeds with the different combinations of defect detection, correction, and subsequent delay.

A process *failure modes and effects analysis* (FMEA) can be applied to the process map to assess whether a high risk of failure is being assumed without an adequate risk mitigation plan. Conversely, if a particular process step has little chance for failure with little impact if it does, then an elaborate inspection step may not be required and could be eliminated. We map the process using an FMEA with a slight difference. The impact on the customer is defined as a time delay to correct the error (failure). It assumes that all errors are eventually detected and corrected.

Conducting a well-planned survey and brainstorming session with the process owners and frontline workers is usually superior to using historical corporate data. The historical data underestimates the defect level and over-estimates the detectability. Much of the rework occurs in the hidden factory and is not tracked by corporate metrics. You are probably engaged in this improvement project because management cannot identify the magnitude or causes of the delays using their existing data.

We will start with the simplified process map for document scanning and indexing of medical claims documents. In reality, there may be some correction of misassigned documents made while the file is being reviewed, but this is infrequent enough that we will not include a rework loop back to the previous step. (Fig. 5.8).

Brainstorm with your team and list the types of errors that occur at each step and record them in a spreadsheet along with the corresponding process step. A cause and effect diagram with the defect "time delay" is a good tool to generate ideas for failure modes at each process step. Each step must have causes of general delay listed. These are the delays that affect all transactions before they enter the process step itself. Any delay within the process will be caused by a specific failure mode and will be listed separately.

The two types of delay will generally have two different strategies for improvement. The general delay owing to the number of items awaiting

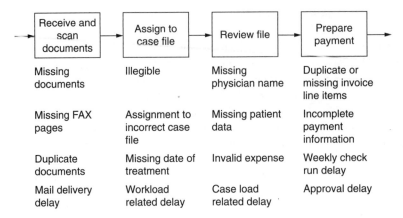

Figure 5.8   Medical Claims Process Map and Failure Modes

processing was shown in Fig. 5.5 as delay, is a problem of understanding the resourcing requirements dictated by customer demand. The delay caused by the failure modes listed in Fig. 5.8 will result in WIP that must be reprocessed. The process must be redesigned or modified to increase the rolled throughput yield to decrease the delay caused by reworking of defects.

With each failure mode, rate the severity of the impact, the occurrence, and detectability using a 10 point scale. The scales for each category are logarithmic and divide the operational range into ten approximately equal intervals. We surveyed the process owners to determine the practical range for values in each category. The typical execution time for scanning or indexing a document is on the order of 10 to 20 seconds. The longest delays caused by errors can be as much as 2 weeks. We have constructed a scale for this project, given typical workloads and processing times (Fig. 5.9).

Rating	Severity (Time Delay)	Occurrence (Likelihood)	Detectability (Certainty)
1	20 s	Every 2 weeks	Every minute
2	1 min	Every week	Every 5 minutes
3	5 min	Twice a week	3 times per hour
4	20 min	Every day	Once an hour
5	1 hr	Every shift	Twice each shift
6	4 hr	Twice each shift	Every shift
7	8 hr	Once an hour	Every day
8	1 day	3 times per hour	Twice a week
9	3 day	Every 5 minutes	Every week
10	7 day	Every minute	Every 2 weeks

Figure 5.9   Scale for Rating Severity, Occurrence, and Detectability

The severity rating is always in terms of impact on the customer. For example, a delay caused by missing information may take your business a total of 10 minutes to make the request for information and 10 minutes to process the missing information when it arrives, but if the entire process takes about 3 days, then this should be rated "9" for severity, not "3" or "4". Each failure mode may also have more than one potential cause. List each potential cause in its own line item. The general delay is defined as the time it would take the first item to be processed after arriving in the queue.

The types of values in the occurrence column are more useful than explicitly specifying probabilities, such as 1:10,000. When we survey process owners, they usually have a good general feeling for the number of times that an error will occur in a given time interval such as an hour, a shift, a day, or a week, but do not express it in terms of an error rate. When rating occurrence for general delay, count the number of transactions that must spend any time in the queue before being processed.

The detectability rating is the one scale which may seem backwards. The explanation is that if an error can occur, but it is not readily detected at the time it occurs, then this has a severe impact on the customer. If a document is assigned to the incorrect case file, then this may not be detected until a payment is made and the customer refiles the claim. Poor detectability usually results in much more work later in the process if it is not detected early. When you are assigning the detectability rating with your team, ask questions in the form of, "If every document had this particular error, how often would you detect it? Once a day? Once an hour? Twice per shift?"

The detectability rating is frequently the hardest to get some agreement from the process owners. Emphasize that the importance is to get at least an estimate of the level of detectability. This is to assess whether to designing a rigorous control plan with mistake proofing is required, or monitoring with a periodic audit is sufficient. For most transactional projects, there is no existing control plan at all. A partial listing of the process efficiency FMEA, for the first few process steps and failure modes, is shown in Fig. 5.10.

Once the ratings for each failure mode are complete, multiply the values for severity, occurrence, and detectability and label the quantity as a *risk priority number* (RPN). A Pareto chart using the RPNs will show the parts of the process with the highest exposure to delay risk.

The Pareto chart in Fig. 5.11 shows that for the *Assign to case file, Review file, and Prepare Payment* process steps, the largest contribution to delay is the clearing of existing backlog. This may indicate that there is a poor understanding of customer demand with subsequent problems with resource allocation.

Process efficiency failure modes and effects analysis (FMEA)									
Process or product Name:	Claims scanning and indexing						Prepared by: A. Perkins		
Responsible	A. Hitchcock						FMEA date (orig)		

Process step/part number	Potential failure mode	Potential failure effects	S E V	Potential causes	O C C	Current controls	D E T	R P N
Receive and scan documents	Missing documents	Slow review	10	—	7	None	4	280
	Missing FAX pages	Slow review	7	—	6	Check header	3	126
	Duplicate documents	Minimal	2	—	5	None	7	70
	Mail delivery delay	Scanning delay	6	—	3	Scheduled	3	54
Assign to case file	Illegible	Rework	3	—	7	Operator check	1	21
	Assignment to incorrect case file	Confidentiality missing information	4	—	6	None	8	192
	Missing date	Misindex	9	—	6	Operator check	5	270
	Workload related delay	Delay	7	—	7	Resource allocation	7	343
Review file	Missing physician name	Delay	9	—	4	Operator check	3	108
	Missing patient data	Delay	9	—	6	Operator check	5	270
	Invalid expense	Wasted resources	2	—	4	Manager check	8	64
	Case load related delay	Delay	10	—	8	Resource allocation	8	640
Prepare payment	Duplicate or missing invoice line items	Incorrect payment	10	—	3	Payment operator check	3	90
	Incomplete payment information	Delay	8	—	4	Payment operator check	2	64
	Weekly check run delay	Delayed payment	9	—	8	Resource planning	3	216
	Approval delay	Delay	8	—	5	None	7	280

Figure 5.10   Process Efficiency FMEA for Medical Claims Processing

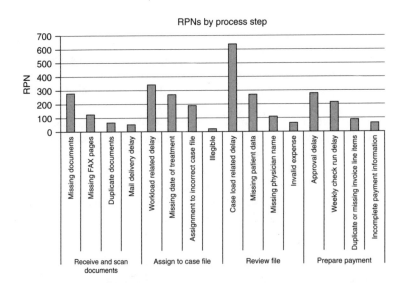

Figure 5.11   A Pareto Chart of RPNs Shows Errors with High Impact on Cycle Time

5.8 Process Indicators (Ys)

Six Sigma emphasizes the relationship, $Y = f(X)$. It mathematically summarizes the fact that the output from a business process is a function of the decisions made by the process owners. We have already seen a number of sources of Ys, the output of the business processes; the Strategic Planning Balanced Scorecard, competitive analysis, customer surveys, and benchmarking are some of the sources of targets for improvement. During the Measure phase you are not concerned with the target for each of these metrics, but rather whether you are gathering the correct data to assess and track them.

The two sets of output indicators come from your two major stakeholders, external customers and internal business leaders. Your *voice of the customer* (VOC) surveys are the source of *critical customer requirements* (CCRs). These are, in turn, the source of your *critical to quality requirements* (CTQs). The strategic planning sessions and Strategic Planning Balanced Scorecard give you *critical to business requirements* (CBRs), which in turn are the source of your *critical to process requirements* (CTPs). The relationship between the VOC surveys and the *critical to success factors* (CTXs) are documented using *quality function deployments* (QFDs) (Fig. 5.12).

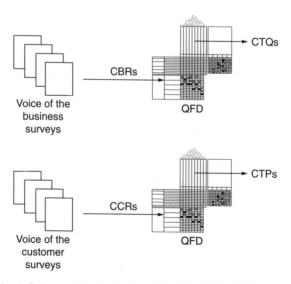

Figure 5.12 A Pair of QFDs Are Used to Distill the Voice of the Customer(VOC) and Voice of the Business (VOB) Surveys to Output Process Variables

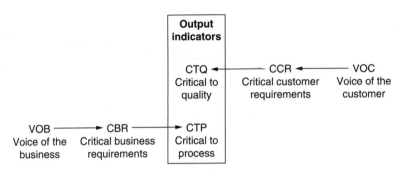

Figure 5.13 The Voice of the Business and the Voice of the Customer Meet to Produce a List of Output Indicators

The set of CTQs and CTPs are the output indicators for your process. These are needed to assess how your business process is performing today, and to evaluate the impact of process improvements you are going to make after the project is complete (Fig. 5.13).

5.9 Data Collection Plan

The list of data to be collected is going to be very large. You have a list of CTPs, CTQs, and a list of Xs from your brainstorming sessions gathered during the process efficiency mapping. You must develop and execute a data collection plan to manage data requirements, sources, roles and responsibilities for team members, and a time line. The most important task is to prioritize your data requirements into an ordered list of decreasing importance. The tool to accomplish this, the *data prioritization matrix* (DPM), is related to the QFD we discussed in Chap. 3 (Fig. 5.14). When you are in this stage, it may seem that you are repeating the work you have already done during the two QFDs associated with the VOC and *voice of business* (VOB) surveys referred to in Fig. 5.12. There are three reasons for conducting a separate data collection plan;

1. Balance the data requirements of the internal and external customers.
2. Minimize overlapping data collection requirements between different stakeholder groups.
3. Assign time lines and roles and responsibilities.

Construct the DPM in a spreadsheet (Fig. 5.14). Begin by listing the CTPs and CTQs in a column down the right hand side. Copy the assigned relative weights from the QFDs as they came from the VOC and VOB surveys that produced them. You will have to balance the entire set of weights for the CTPs versus the CTQs. Check to make sure you have used most of the scale of 1 to 10 when ranking the CTXs. This does not mean that the "10" is ten times more important than the "1". The tool works best at discriminating

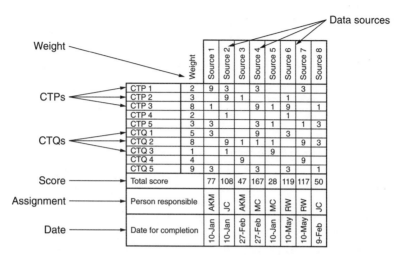

	Weight	Source 1	Source 2	Source 3	Source 4	Source 5	Source 6	Source 7	Source 8
CTP 1	2	9	3		3			3	
CTP 2	3		9	1			1		
CTP 3	8	1			9	1	9		1
CTP 4	2		1				1		
CTP 5	3	3			3	1		1	3
CTQ 1	5	3			9		3		
CTQ 2	8		9	1	1	1		9	3
CTQ 3	1			1		9			
CTQ 4	4			9			9		
CTQ 5	9	3			3		3		1
Total score		77	108	47	167	28	119	117	50
Person responsible		AKM	JC	AKM	MC	MC	RW	RW	JC
Date for completion		10-Jan	10-Jan	27-Feb	27-Feb	10-Jan	10-May	10-May	9-Feb

Figure 5.14 Data Prioritization Matrix-Assign Weights to CTPs and CTQs, Populate the Matrix with Relationships (9,3,1,0-High, Medium, Low, None) Total the Scores, then Assign Tasks and Dates for Completion

and comparing values using a relative scale, not an absolute one. List the potential data sources across the top row. These come from brainstorming the requirements for planned data collection or mining existing or historical data.

The center of the matrix requires a large number of decisions on the part of the team. These values indicate the strength of the relationship between the CTX and the potential data source. Address the relationships one at a time and limit discussion to 20 seconds or less. Anything requiring more debate is put "on hold," for later discussion. Voting using a show of fingers for *high*, *medium*, *low*, and *none* helps make the session run quickly and prevents one person from dominating the results.

When the team has populated the matrix, check for blank columns. These indicate potential sources of data that have no relationship with the CTXs. Do not collect or assemble this data, even if it is easily available. Blank rows present a bigger problem: you have an issue that is important to one of your stakeholders, but no data source that can shed light on the situation. You must come up with a source of data to address this deficiency. This will require adding a data design and collection component into the overall data collection plan.

Calculate the scores for each data source by multiplying each CTX weight by its corresponding cell value and summing the subtotals for each column (Fig. 5.14). The score for the first data source is:

$$(2 \times 9) + (8 \times 1) + (3 \times 3) + (5 \times 3) + (9 \times 3) = 77 \qquad (5.6)$$

Prepare a Pareto chart of data collection priorities using the scores from the DPM. The highest scores come from data sources that have strong relationships with the most important CTXs. The team can now balance the need for gathering relevant data against reasonable requirements for timeliness and effort by choosing a cutoff point in the Pareto chart. Assign data collection task to team members and complete the DPM with expected dates for completion.

5.10 Cycle Time, Execution Time, and Delay Time

You will be measuring a large amount of different types of time data. It is important to collect the data in as "raw" a form as possible. It is common to have access to historical data for signoff times in transactional projects, but it is our experience that people are very good at documenting when they completed a task, but usually not as diligent at recording when they began, therefore, delay time for process steps cannot be calculated. Do not assume that a successive task begins as soon as the previous step is completed. The calculation of such quantities as interarrival times, variance from customer want dates, execution times, and so on, can be done after the fact (Fig. 5.15).

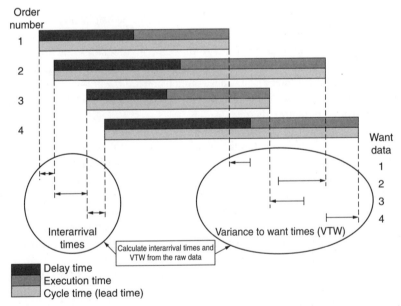

Figure 5.15   Collect the Raw Data Necessary to Calculate Time Related Metrics

	A	B	C
1	**Field Name**	**N=1**	**N=2**
2	Order(n)	5123345	5123346
3	OrderEntryDate(n)	Jun-12-2005 7:33 PM	Jun-18-2005 6:26 AM
4	PromiseDeliveryDate(n)	Jun-30-2005 3:57 PM	Jun-30-2005 2:50 AM
5	StartExecutionDate(n)	Jun-28-2005 6:21 AM	Jun-27-2005 5:14 PM
6	StopExecutionDate(n)	Jul-02-2005 3:34 PM	Jun-28-2005 10:02 PM
7	**CalculatedFields**		
8	DelayTime(n)	15.45	9.45
9	ExecutionTime(n)	4.38	1.20
10	CycleTime(n)	19.83	10.65
11	VarianceToWantDate(n)	9.18	−1.20
12	InterarrivalTime(n)		5.45
13	**CalculatedFields**		
14	DelayTime(n)	=B5–B3	=C5–C3
15	ExecutionTime(n)	=B6–B5	=C6–C5
16	CycleTime(n)	=B8+B9	=C8+C9
17	VarianceToWantDate(n)	=B6–B4	=C6–C4
19	InterarrivalTime(n)		=C3–B3

Figure 5.16   Four Data Fields for Each Process Step Will Allow Calculation of Process Metrics

Figure 5.16 shows some example data for two successive transactions. Rows 14 to 18 show the formulae for the calculation of the metrics displayed in rows 8 to 12. The convention for calculating variance from customer want date is that negative values indicate the transaction was delivered earlier than the customer want date.

If you are at a stage where you are designing your data collection plan in the absence of any useful historical data, you should record the start and stop times for each individual process step. The time between one task finishing and the next one starting is the delay time for the second step. It is usual to associate the delay step with the subsequent execution step as shown in Fig. 5.15.

5.11 Data Types

Traditional Six Sigma projects will have a variety of different types of data. Discrete defect data may be simply counted and enumerated and expressed as a percentage of the total. Continuous defect data from a physical measurement, such as the diameter of a hole or a duration of time, can be recorded and compared against specification limits. Transactional lean

Six Sigma projects tend to have a much wider variety of data types and it is important to classify data to assess the confidence you may place on them. Customer satisfaction data may be recorded as simply high, medium, or low.

Data can be considered as lying along a scale ranging from highly detailed to coarse (Fig. 5.17). If it is at all possible, try to redefine your operational definitions to move toward more detailed data types. Imagine that you have a project where you are recording when a shipment has arrived into a facility. Do not record the shipments as early, on time, and late (ordered category data) if you have access to reliable information about arrival times at the level of hour, minutes, and seconds (ratio data). Rather than recording customer satisfaction data as high, medium, or low (few ordered categories), it is better to record as a response from 1 to 10 (many ordered categories/many counts).

5.11.1 Discrete or Attribute Data

Discrete data cannot be divided into smaller increments and must occur in fixed values. An easy way to tell if you have discrete data or not is to think about how you would take an average value—if it has no meaning or could not occur, then your data is discrete. What would be the meaning of an average gender for a customer base? Could an application have 1.43 errors? Given eastern, central, southern, and western sales regions, what would be the "average" sales region-Kansas?

5.11.2 Binary

As the name implies, there are only two possible values—pass/fail, on/off, yes/no, high/low, defect/non-defect, and so on. If you encounter this data type, try and see if there is a way of quantifying the degree of the pass/fail.

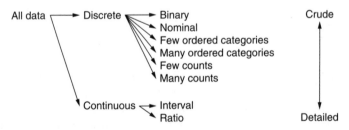

Figure 5.17   Data Types Range From Crude to Detailed

For example, when examining application forms, rather than classifying the form as complete or incomplete (binary data), see if you can count the number of occurrences of missing information on the form (count data).

5.11.3 Nominal

This type of data is commonly found when summarizing other data when the "bins" correspond to categories, but the categories have no particular order. Analyzing sales figures (continuous data) broken out by geographical region (nominal data) illustrates this data type.

5.11.4 Few Categories/Many Categories

Everyone can agree that this type of data can be placed in order from one extreme to the other, but the differences between adjacent values are meaningless—the difference between *very satisfied* and *satisfied* for a survey response is not the same as the difference between *dissatisfied* and *very dissatisfied*. It is common, though not strictly valid, to assume that the intervals between successive values are approximately equal and place the categories on a numerical scale. Do not think that the conversion has transformed your data from categorical to count data; this is only to aid in summarizing responses. As it is with count data, as the number of categories increases, the discrimination power of the data increases—letter grades have more discrimination power than a pass/fail mark (Fig. 5.18).

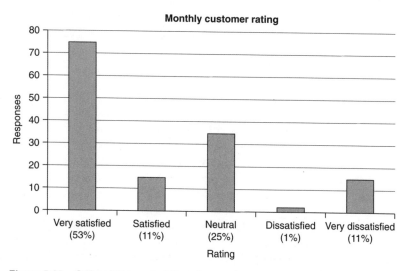

Figure 5.18 Ordered Categorical Data Summarized in This Manner Might Imply That the X-Axis Is Numerical Data

5.11.5 Many Counts/Few Counts

Your project may be about quantifying errors on documents. There is clearly a difference in the process performance when a document has one, two, three, or more errors and your data should be able to capture that difference. As the number of defects increases, the discrimination power increases accordingly. There is fine difference between a case claim file that contains 10 errors versus one that contains 12. The amount of detail in the data increases as the number of possible values of defect counts increases.

When the probability of a defect is independent of another occurrence in the same unit, the probability of different numbers of defects occurring is described by the Poisson distribution, where the distribution of the number of occurrences, X, of some random event is given by

$$\Pr(X = x) = \frac{e^{\lambda} \lambda^x}{x!} \qquad x = 0, 1 \ldots \tag{5.7}$$

It has a single parameter, the average number of *defects per unit* (DPU).

$$\text{DPU} = \frac{\text{number of defects}}{\text{number of units}} \tag{5.8}$$

An important quantity derived from Equation 5.7 is the probability of zero defects, also called the rolled throughput yield. The simplified equation for zero defects is,

$$\Pr(0) = e^{-\text{DPU}} \tag{5.9}$$

We had a project looking at errors of all kinds for documents in case files. For the set of data shown in Fig. 5.19, the total number of errors was 757 and the total number of case files was 150. Equation 5.8 gives a DPU of 5.05 and Equation 5.9 gives the probability of a case file having zero defects as 0.64 percent. Expressed as a *defects per million opportunities* (DPMO), the error rate was 993,600.

The data shown in this example was from the Measure phase of the project. During the Analyze phase we counted the number of errors of different kinds for different document types in order to determine the root causes of defects. Remember about the unit of transaction at the customer level. (See Section 5.1)

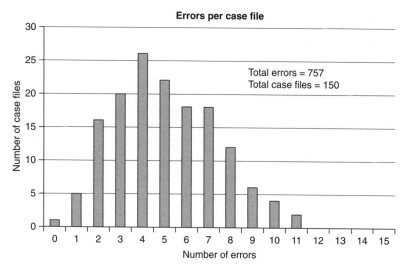

Figure 5.19   Examples of Count Data (Many and Few)

5.11.6 Continuous or Variable Data

These two names are equivalent and tend to follow whether the context is statistical analysis or *statistical process control* (SPC) respectively. If your data is variable, this means that in the range that is relevant to the application, all values are possible and could, in principle, be measured to an arbitrarily large number of decimal places. Time will be the most common continuous variable in your project. You may be measuring elapsed time in hours and minutes, but there is no reason why you could not increase detail and measure in seconds and fractions of seconds. Money can be measured to a large number of decimal places so that it can always be treated as a continuous variable.

Whether the variable data is ratio or interval depends on the meaning of a value of zero. If it corresponds to the complete absence of something then the data is ratio, variable data. Examples of ratio data are length measured in millimeters or elapsed time in seconds. If the value of zero does not correspond to an absence of the quantity or an arbitrary value, then the data is interval, variable data. Examples of interval data are temperature measured in degrees Fahrenheit or date/time stamps on the Julian calendar.

5.11.7 Pseudocontinuous Data

Statisticians have a few basic rules for dealing with the restrictions of different data types. One example we have seen so far is assigning numerical values to the categories of ordered categorical data of a customer satisfaction scale. The most important rule is the point at which count data can be treated as continuous data for the purposes of calculating process capability and conducting statistical tests.

There will be situations when the data is derived from a continuous source, but has been "binned" into categories for ease of data collection. In these cases, if the number of bins between the upper and lower specification limits is about ten or greater, then the data can be treated as continuous data. You can summarize the data using the mean, standard deviation, kurtosis, and so on. The distribution does not need to be symmetric or bell shaped.

Figure 5.19 shows a distribution of counts of defects per case file. It is not too great a leap of faith to imagine a smooth curve that could be used to describe the defect rate. The discrete, Poisson distribution can be approximated by a normal distribution when $n(p)$ or $n(1-q)$ is greater than about five. For example, if we were summarizing survey data where 5 percent (p) of the 250 people (n) surveyed answered 'yes,' to a particular question, $250(.05) = 12.5$ is greater than 5 and the distribution of responses from the population can be summarized using the normal approximation.

5.12 Probability Distributions

Continuous data will tend to follow some common probability distributions. Cycle time data is usually skewed to the right and is limited to positive values. Production data tends to be skewed to the left and has a large hump near the production maximum. Variance to customer delivery want date data tends to be spread very wide with a sharp narrow peak near the promise date. Customer interarrival times, delay times, and process execution times tend to follow a few well-understood distributions. Most of the data you will see in your project is limited to a few different types.

Statistical software packages have the capability to calculate probability plots to determine the distribution that best fits your data. Your observed data is used to calculate the parameters of the distribution you are testing, and then a plot of theoretical percentiles against observed percentiles is constructed. If the plot is a straight line, then there is a good fit between the observed and theoretical probability distributions. In general, a steep line indicates a narrow part of the distribution and a shallow line indicates a wide part of the distribution.

A set of data was constructed for the cycle time of a business process that consists of two parallel streams, each handling 50 percent of the input. One work stream takes an average of 10 minutes, while the other takes an average of 20 minutes (Fig. 5.20).

The histogram and probability plot for the combined set of data are shown in Fig. 5.21. The histogram shows a spike near the 10 minute mark and a low, long hump that extends out beyond 30 minutes. The average is about 15 minutes, but the probability plot and histogram indicate that the process is not a symmetric bell-shaped curve. The initial part of the probability plot is quite steep from about 7 to 12 minutes. This corresponds to the execution time for the fast process. At about the 50th percentile, and after about 12 minutes, the probability plot flattens out, corresponding to the slow process.

The conclusions from the probability plot are:

- The process cannot be described by a single, normally distributed process.
- The process consists of an initial, fast, and consistent process and a slower one.
- The long execution times come from only one component of the process.
- The ratio of the fast to the slow process is about 50:50.
- An upper specification limit of 15 minutes would be exceeded by about 100 percent of the applications for new customers and about zero for returning customers.

5.12.1 Execution Time Data

During a project on application processing, the time between receiving an application and having the application sorted and classified was tracked. The time stamp for "receipt" was the time that the sorting process actually started and did not include any delay time as shown in Fig. 5.15. The

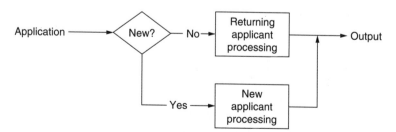

Figure 5.20   A Fast and a Slow Work Stream in Parallel

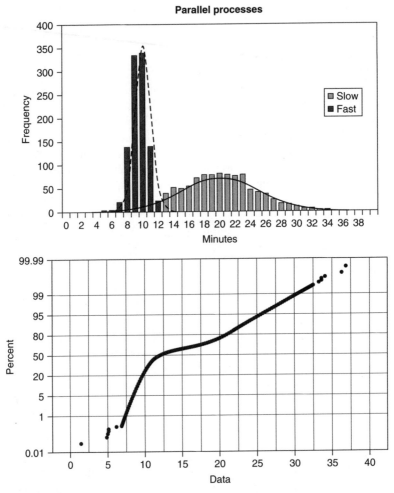

Figure 5.21 Histogram and Probability Plot of Overlapping Fast and Slow Processes

histogram and probability plot of "timeliness" are shown in Fig. 5.22. The software will assume a normal (Gaussian) distribution by default to fit a curve to the observed data.

The histogram in the upper part of the figure shows a poor fit between the bell-shaped, symmetric, Gaussian distribution. The left hand side of the distribution is narrow, while the right hand side is skewed out to extreme values. The probability plot below shows a bent line, indicating a poor fit between the observed

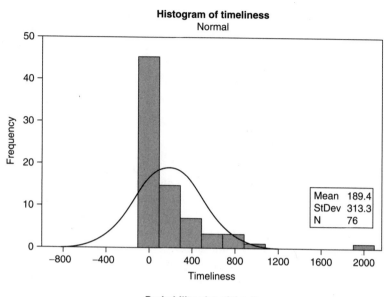

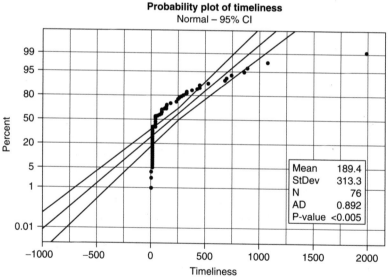

Figure 5.22   Timeliness Data Does Not Fit the Gaussian Distribution

data and a Gaussian distribution with a mean of 189.4 hours and a standard deviation of 313.3 hours. The software has also run a "goodness of fit" test for the observed against the theoretical distribution. The p-value of less than 0.005 indicates that the fit between the two is poor enough to violate the test.

The probability plot itself indicates reasons why this test failed. The left side of the probability plot is steep, mirroring the narrow distribution at short times, while the rest of the probability plot is shallow, indicating the broad distribution to longer times. This data might be described better by a probability distribution other than the bell-shaped distribution in Fig. 5.22.

5.12.2 The Normal or Gaussian Distribution

This distribution is a commonly encountered distribution and is often observed when physical measurements of some kind are taken. It shows a familiar symmetric, bell shape. We have observed the normal distribution for the heights of people entering a bank, the weight of product in a can, the execution times for some manufacturing processes, and the number of customers entering a government office on each Tuesday of the month. The normal distribution is not usually seen when measuring time dependent data such as cycle time, delay time, or execution time. It is characterized by two parameters, the location (μ) and the scale (σ) using the equation,

$$f(x) = \frac{1}{\sigma\sqrt{2\pi}} \exp\left(-\frac{(x-\mu)^2}{2\sigma^2}\right) \tag{5.10}$$

Examples of data following this distribution are shown in Fig. 5.20.

5.12.3 The Exponential Distribution

Customers, orders, and requests do not often arrive with any regularity. When the probability of a customer arriving is independent of another customer arriving, then interarrival time between customers follows an exponential distribution. Exponential interarrival times are often seen for incoming calls to customer service centers. It has only the single parameter—the average time between customers. It is quite high at low values, and stretches out to the right. Some statistical references will specify the distribution using the *scale* parameter, b, where $b = 1/\lambda$.

$$f(x) = \lambda e^{-\lambda x}, x > 0 \tag{5.11}$$

You may have to calculate the average time between customers from data expressed as the number of customers that arrived in a given time interval. This is also referred to as "pitch" in the lean literature. By grouping the customer demand into packages, the *takt time* (pace of customer demand) can be smoothed to manage the large variation in customer demand. In a business process, it is a way of providing a fairly consistent group of tasks to downstream processes.

There is always a danger that the smoothing effect of grouping customer demand can mask quick changes in customer demand. Figure 5.23 summarizes the number of patients arriving hourly at the triage desk of a walk-in medical clinic. The doors of the clinic open at 6:30 a.m. and close at 8:00 p.m. This is a picture of the "pitch" of the customer demand.

The average interarrival time for the entire day is 810 minutes for 333 patients (2.43 min/patient), but the figure shows that the number of patients arriving per hour changes throughout the day. It would be a mistake to make the pitch equal to 333 patients per 13.5 hour interval. The interarrival time of customers would best be described by a series of exponential distributions where the average time between patients changes hourly during the day.

We can test how well this assumption is followed by examining the interarrival times for the thirty seven patients who arrived in the one hour interval between 9:30 and 10:30 in the morning. The average patient arrival rate during that hour was 1.62 minutes and the interarrival times were skewed to the right (Fig. 5.24).

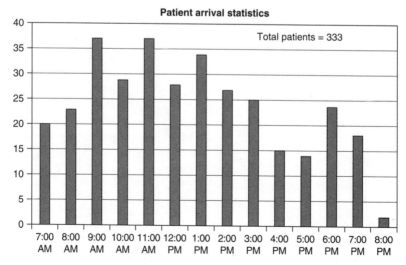

Figure 5.23 Patient Arrival Time Statistics

Summary for 8:30–9:30 interarrival time (min)

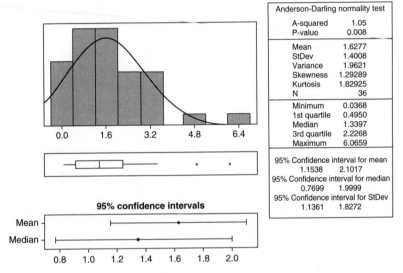

Figure 5.24   Descriptive Statistics of Patient Interarrival Times

The probability plot of the interarrival times confirms an exponential distribution with a mean of 1.62 minutes per patient (Fig. 5.25). The "goodness of fit" test shows a p-value of 0.808 (greater than 0.05), indicating the exponential distribution fits the sample of data for the full hour.

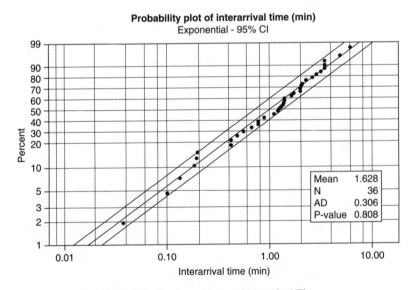

Figure 5.25   Probability Distribution of Patient Interarrival Times

5.12.4 The Weibull Distribution

When someone is processing applications from a large pool in their "in" box and selects the applications at random, then the amount of time the application spends in the "in" box is not dependent on the length of time it has waited. The waiting time will show an exponential distribution of delay times.

Processes do not always have this independence of between events. The Weibull distribution was first used to allow for the time dependence of product failure rates. The distribution is described by three parameters, shape (a), scale (b), and threshold (θ) using,

$$(x) = \frac{a(x-\theta)^{a-1}}{b^a} \exp\left(-\left(\frac{x-\theta}{b}\right)^a\right)$$ (5.12)

The shape parameter indicates the time-dependence of the function

- When a product wears out, it becomes more and more likely to fail as time passes and $a > 1$.
- If customer churn, or product failure rates do not change as time passes, then $a = 1$.
- When customers become more indebted to the customer loyalty program as time passes, and they become less and less likely to defect to a competitor, then $a < 1$.

When $a = 1$ is substituted in Equation 5.12 and simplified, the distribution becomes Equation 5.11 with $\lambda = 1/b$, showing that the exponential distribution is a special case of the Weibull distribution. The Weibull distribution becomes approximately Gaussian when $a = 3.4$.

We had a project where we gathered data on the time that sales staff was spending on a variety of tasks. They are least productive when they are spending a lot of time performing administrative tasks. Figure 5.26 summarizes the amount of administrative time that 97 sales staff were spending during one week.

The majority of sales staff were spending a short amount of time on administration. While the average was about 4.5 hours, there were some people spending 45 hours per week! The probability plot in Fig. 5.27 shows a good visual fit with a Weibull distribution. The shape factor of 0.62 shows that the time spend on administration tends to decrease as the staff spend more time on it—there is some pressure to spend less time on it. The p-value of 0.020 indicates that it is not a perfect fit to a Weibull, but is still much better than assuming a Gaussian distribution.

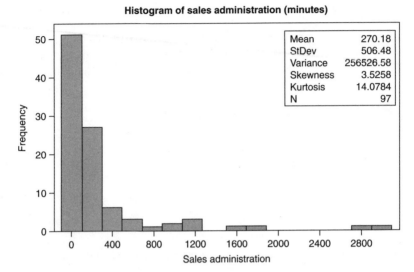

Figure 5.26   Sales Administration Time Per Week (Minutes)

5.12.5   The Log-Normal Distribution

When the result of a large number of random events is additive, the result is the Gaussian distribution. When the result of a large number of random events is multiplicative, the result is the *log-normal distribution*. It is skewed

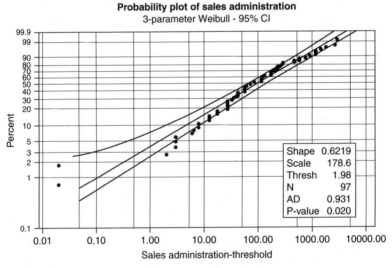

Figure 5.27   Probability Plot of Time Spent on Administration by Sales Staff

to the right and is commonly observed for financial data. Sales data have a large number of small transactions with a small number of very large ones. The distribution of items in a warehouse also will follow a log-normal distribution. It is common enough in financial data that it forms the basis for a fraud detection technique known as Bedford's law that relies on digit frequency analysis. The equation for the distribution is similar to that used for the normal, or Gaussian distribution (Equation 5.10) and is characterized by three parameters, the scale (ζ), the location (σ), and the threshold (θ),

$$f(x) = \frac{1}{(x-\theta)\sigma\sqrt{2\pi}} \exp\left(-\frac{\ln((x-\theta)-\zeta))^2}{2\sigma^2} \right) \qquad (5.13)$$

During a year end audit of the items in a large warehouse, we examined the dollar value of each item in the warehouse. There were many small items and only a few very large items. The items ranged in cost from $0.01 to $4,200,000 with an average cost of $21,000. Figure 5.28 shows the data of a log-normal distribution.

This distribution shows the importance of tracking the actual cost of inventory or WIP in your lean Six Sigma project. Merely using an item count and an average inventory item cost will grossly underestimate the impact of inventory on cash flow.

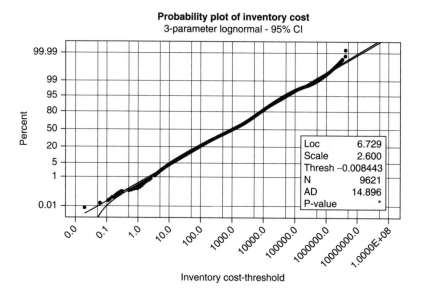

Figure 5.28   Log-Normal Distribution of Warehouse Item Costs

5.13 Data Distributions from a Medical Clinic

When you are gathering data, conducting calculations, or simulating processes, you must be able to characterize the identities and parameters of the probability distributions you are going to use. We have assembled a sample of data gathered during a lean Six Sigma project at a private medical facility operating within a hospital.

Blood and other biological samples are collected from patients around the hospital and sent to the centralized laboratory for processing. Costs depend on a large variety of factors, and are defined in the service contract. Some samples require isolation from other samples while some samples are "stat" (urgent). Data have been collected for a sample of some of the process steps involved and are summarized in Fig. 5.29.

In order to describe these distributions, Minitab (See Appendix C) was used to determine the type of distributions for each set. The *distribution analysis* option generates a series of probability plots with statistical tests assessing goodness of fit for a large number of distributions.

Lab time for isolation sample	Lab travel time for stat isolation sample	Collection travel time for stat sample	Revenue per claim
777	129	110	154
826	19	22	95
388	6	185	194
658	2	183	92
607	33	3	401
518	9	20	78
666	68	13	108
718	15	106	180
822	91	45	99
826	5	4	227
592	18	13	104
600	16	58	114
673	28	92	184
487	42	169	110
733	108	83	88
746	100	35	83
773	32	102	183
604	49	36	129
437	191	99	130
748	33	268	239

Figure 5.29   Time and Revenue Data for Laboratory Services

5.13.1 Lab Time for Isolation

Isolation samples spend some time in the lab. The distribution ID plots show that the data could be described quite well by a few different probability distributions (Fig. 5.30).

We have chosen the normal distribution with a mean of 660 minutes and a standard deviation of 129.7 minutes. The p-value for goodness of fit test is 0.379, indicating an acceptable fit (Fig. 5.31).

5.13.2 Lab Travel Time for Stat Isolation Sample

Once samples are collected, they spend a certain amount of time moving between the different areas within the lab. The distribution ID plots indicate these travel times fit the exponential distribution. Since the Weibull is a superset of the exponential, the Weibull also fits the data quite well (Fig. 5.32).

We have chosen to fit the data with an exponential distribution with a mean time of 49 minutes. The p-value for the goodness of fit test is 0.963 (Fig. 5.33).

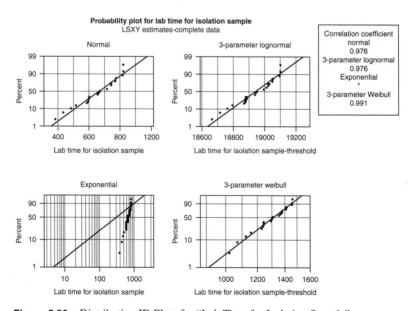

Figure 5.30   Distribution ID Plots for "Lab Time for Isolation Sample"

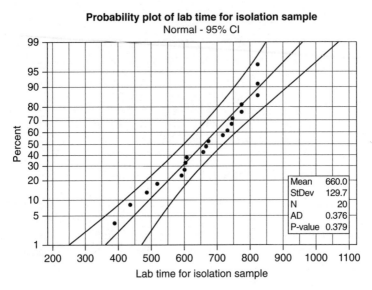

Figure 5.31   Normal Probability Plot for "Lab Time for Isolation Sample"

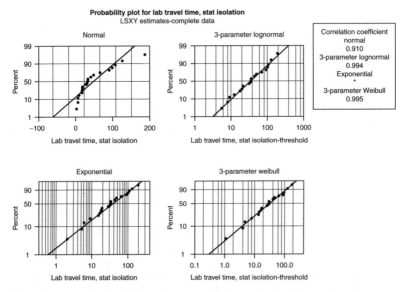

Figure 5.32   Distribution ID Plots for "Lab Travel Time for Stat Isolation Sample"

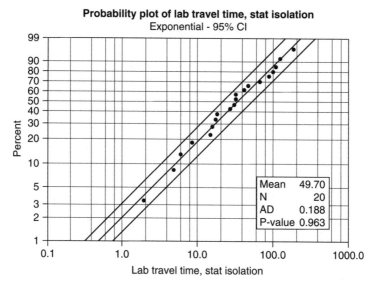

Figure 5.33   Exponential Probability Plot for "Lab Travel Time for Stat Isolation Sample"

5.13.3 Collection Travel Time for Stat Sample

When a request is made for a sample to be taken, a dispatch is sent to the staff. This is the time it takes the staff member to physically travel to the patient to obtain the sample (Fig. 5.34).

The Weibull distribution was chosen to investigate the shape parameter. If a > 1, then the sense of urgency increases as the request becomes older (Fig. 5.35).

The shape parameter of the distribution is 1.062, indicating that the distribution is essentially exponential. The sense of urgency does not increase as the request becomes older. This data could also be described as an exponential distribution with a mean of 82.3 minutes (data not shown).

5.13.4 Revenue Per Claim

A sample of 20 claims from historical financial records was obtained and is listed in Fig. 5.29. The distribution ID plot shows a good fit with a lognormal distribution (Fig. 5.36).

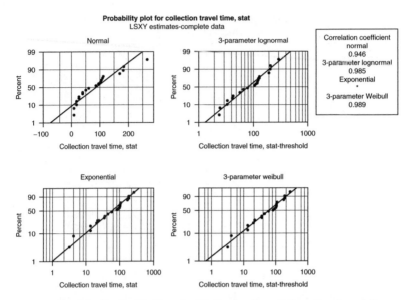

Figure 5.34   Distribution ID Plots for "Collection Travel Time, Stat Sample"

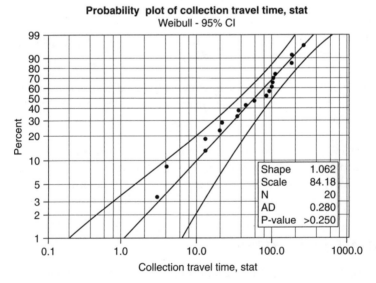

Figure 5.35   Weibull Probability Plot for "Collection Travel Time, Stat Sample"

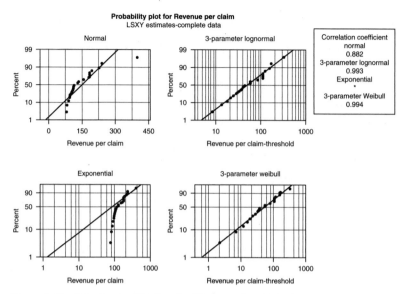

Figure 5.36   Distribution ID Plots for "Revenue Per Claim"

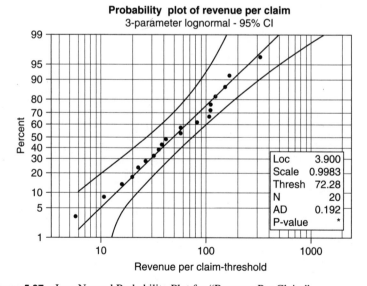

Figure 5.37   Log-Normal Probability Plot for "Revenue Per Claim"

The log-normal distribution makes more logical sense than the time-dependent Weibull distribution and is consistent with other examples of financial data. The distribution has a threshold of $72.28 reflecting the minimum, fixed processing charge (Fig. 5.37).

5.14   Process Capability for Cycle Time

Traditional Six Sigma projects require the definition and measurement of a project "defect." If your project is focused on cycle time reduction, then your will have *upper specification limit* (USL) for the cycle time for individual process steps. These limits may have come from competitive benchmarking studies, VOC surveys, or the company's strategic plan.

In these cases, you can calculate the proportion of defects that are expected to occur, given the random sample of data you have gathered. The usual assumption is that the data follows a particular, well-understood probability distribution. Statistical software can be used to calculate the defect proportion. This is usually expressed as a DPMO level.

The cycle time for an accounts payable process is shown in Fig. 5.38. The USL was 60 days and the cycle time approximately followed a Weibull distribution. This information was used to calculate the number of times that an account would be expected to take longer than the 60 day limit (DPMO = 232,130).

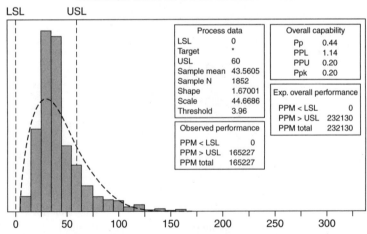

Figure 5.38   Process Capability Using a Weibull Distribution

In some cases the data may require a mathematical transformation to force it to follow a probability distribution allowed by the software. If this is the case, note the transformation and apply it to the USL and *lower specification limit* (LSL) before proceeding with the process capability calculation.

5.15   Hazard Plots for Cycle Time

The discussion about the parameters of the Weibull distribution showed the value of knowing the time dependence of the probability function. Whether the probability of an event occurring decreases, stays the same, or increases with time has a bearing on the evaluation of the process. Effective customer retention programs should result in customers having a lower likelihood of leaving as they stay longer. The escalation process for the collections function in finance should result in customers being more likely to pay as the accounts become more overdue.

The hazard function evaluates this likelihood by the calculation of the instantaneous failure rate for each time, *t*. For products, the failure rate is usually high to begin with, low in the middle, and high again at the end of the plot. The initial stage is often called the infant mortality stage, the middle section is the normal life stage, and the end of the curve, where the failure rate increases again, is the wear out stage. This behavior is frequently called a "bathtub" function.

The parametric method to calculate the hazard function begins by fitting a probability function to the data, then calculating the theoretical function based on the probability distribution. This results in a smooth hazard function, but is limited to a set of well-understood distributions (Fig. 5.39). This method would work well in the case of "lab time for isolation samples" (Fig. 5.30 and Fig. 5.31), because the distribution fits the normal distribution quite well.

As we saw in Section 5.13, the Weibull is quite general and extremely useful in describing time dependent data. The shape parameter (*a*), indicates whether the instances become less likely to occur as they become older (wearing in), more likely to occur as they get older (wearing out), or show no time dependence (random failure). Three data sets were constructed using the Weibull distribution and the parameters in Fig. 5.40. The hazard plots show that the instantaneous failure rates either increase (shape = 1.04), decrease (shape = 0.91) ,or remain approximately constant (shape = 1.00) as time passes.

In a case where a single distribution does not fit the data very well, the nonparametric method is a better choice for calculating the hazard function. The instantaneous failure rate is calculated for each individual point when a

Parametric hazard plot for lab time for isolation samples
Normal
complete data - LSXY estimates

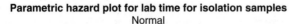

Table of statistics	
Mean	659.95
StDev	133.749
Median	659.95
IQR	180.424
Failure	20
Censor	0
AD*	0.940
Correlation	0.976

Lab time for isolation samples

Figure 5.39   Hazard Plot for "Lab Time for Isolation Samples" Shows the Rate Starts Very Low and Increases as the Samples Get Older (Parametric Method)

Parametric hazard plot for weibull processes
Weibull
complete data - LSXY estimates

Group		Table of statistics				
——	Weib 0.9	Shape	Scale	Corr	F	C
- - -	Weib 1.0	0.91214	10.0406	0.998	1000	0
·······	Weib 1.1	0.99578	10.0000	0.999	1000	0
		1.04327	9.9824	0.997	1000	0

Time

Figure 5.40   Probability of Failure as a Function of Time for Different Weibull Distributions (Parametric Method)

failure occurs. Since the rate is calculated for each point in the data set, the graphs are not as smooth as those using the parametric method.

The following sets of data are from an accounts payable project executed at a large service center. It is the same project discussed at the beginning of this chapter (Fig. 5.1). The "days to pay" are the number of calendar days after the invoice was mailed to the customer. The standard terms for payment are *net +30 days* (Fig. 5.41).

The hazard plot indicates a fairly low rate of payment (failure), beginning at the shortest payment made four days after mailing the invoice. The rate remains constant until about day 32 or so, when it starts to increase at a constant rate each day. After about 62 days, the increase in rate of payment changes again. The graph indicates that there is an effective, time-dependent penalty invoked after 30 days followed by a harsher penalty after 60 days. The underlying business process in this project was that the accounts receivable would handle accounts less than 30 days overdue before transferring the accounts to collections when they became older.

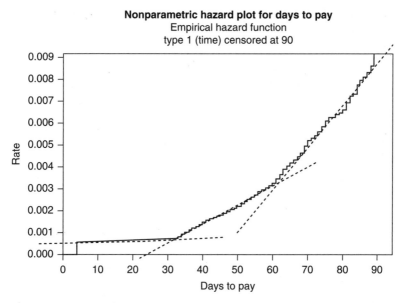

Figure 5.41 Hazard Plot of "Days to Pay" for Accounts Receivable (Nonparametric Method)

We could not find a single probability distribution to describe the data. The hazard plot indicates there are three successive, time dependent processes that best describe the customer payment behavior for payments made between four and ninety days after mailing the invoice.

Another example of a time-dependent process is the generation of quotes for customers. We executed a lean Six Sigma project aimed at decreasing the span in variance to customer want date for the process of generating quotes in a large equipment manufacturing company. This process can take a very short period of time for routine small orders to weeks or months for large installations. The customer information was gathered and entered into an order entry system and the quote was divided into parts and services components. Parts were identified and defined (drawing numbers), service levels were defined and priced, price and cycle time for parts were determined by the sourcing and engineering staff, then the documents were assembled and sent to the customer (Fig. 5.42).

This is a business process where each order is different, but follows roughly the same process. Unlike the examples for lab time for isolation samples or accounts payable, each quote has a customer want date associated with it. Some quotes are required (and promised) immediately for an unplanned breakdown, while some are required for stocking spare parts and need to be delivered only when yearly budget planning is taking place. Note that the want date for the quote is separate from want dates for service and parts components of the quote.

We gathered cycle time data for quote generation for 40,602 requests over a full calendar year. The distribution of cycle times was quite unusual (Fig. 5.43). The data ranged from three to 550 days. The distribution was highly skewed to the right with 30 percent of the quotes being generated in three days or less, and about 23 percent of quotes taking longer than 28 days.

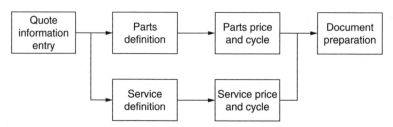

Figure 5.42   Quote Generation Process for Manufacturing Large Equipment

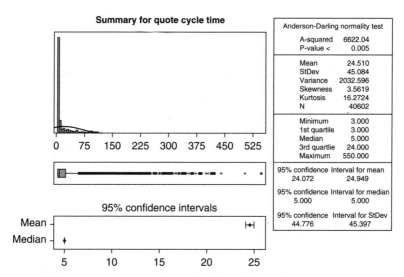

Figure 5.43   Summary of Cycle Times for Quote Generation

The hazard plot showed that the rate of execution did not change as the requests aged beyond about 10 days. After about that time, the rate of execution was time-independent, as you might expect if requests were handled in no particular order as they got older (Fig. 5.44).

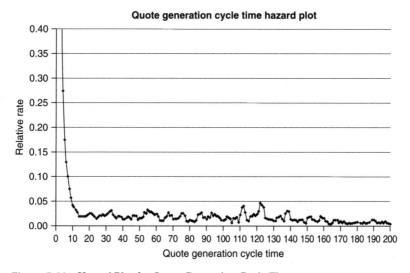

Figure 5.44   Hazard Plot for Quote Generation Cycle Time

5.16 Summarizing Defect Levels
with Customer Want Dates

The previous examples show the characterization of cycle time with fixed USL and LSL common to all transactions. A project where cycle time reduction is the single focus will use a standard upper cycle time limit and apply it in all cases. This may make sense where all accounts are expected to be paid in 30 days, however, many transactional processes show a great range of cycle times owing to the variation of the entities passing through. The number of line items on different orders will vary greatly. It would be unrealistic to impose a common USL on the cycle time for large and small orders.

When transactions are complex and take some time to execute, the customer service representative or salesperson will ask the customers for their time requirement, consider their company's capacity, and make a promise to the customers. In this sense each transaction will have its own individual specification limit. This promise may become part of the terms and conditions section of a service contract. This approach of individual specification limits with each transaction is consistent with flexible service offerings and customer focus.

We will re-examine the data for quote generation using the individual want dates for each transaction. The *variance to want* (VTW) measurement for all transactions is defined as:

$$(\text{Quote\_delivery\_date}) - (\text{quote\_want\_date}) = \text{VTW} \qquad (5.14)$$

as shown in Fig. 5.16. Quotes delivered early are negative and quotes delivered late are positive. The ideal case is a VTW of zero. Any variation from zero is considered as a defect by the customer. Early deliveries are just as inconvenient for the customer as late deliveries. Any variation in VTW indicates inconsistent execution time, poor understanding of company capacity, or poor order management. VTW data never follow easily described distributions and are difficult to measure and track in terms of DPMO or other common measures of process capability. The summary of the VTW data for the quote generation process is shown in Fig. 5.45.

The VTW data shows a peak near the customer want date with very broad shoulders to the left (early) and to the right (late). It does not follow the calculated normal curve.

The defect calculation defined in traditional Six Sigma projects uses the same USL and LSL as specified by the customer for all transactions. A probability distribution is fitted to the data and combined with the specification limits— calculate the proportion of defective transactions.

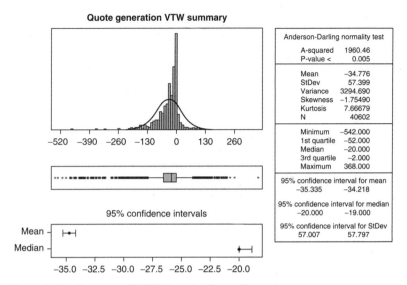

Figure 5.45   Summary of VTW Data for Quote Generation

The data sample is usually limited and a good fit between the data and the distribution is necessary. Commonly assumed distributions will underestimate the extreme values in the tail regions of commonly found VTW data.

Transactional projects usually have enough data on VTW that fitting a theoretical distribution to the data is neither acceptable nor necessary. USLs and LSLs enter the calculation as individual need dates obtained from the customer for each transaction. Summarizing the data by tabulating the percentage of data points exceeding a corporate limit of variance to customer want date can be done, but it is a clumsy metric to report and means little to the customers.

It is best to choose the percentiles and then track and report the width of the distribution. The 5th and 95th percentiles are calculated using the VTW data and reported. The span is the difference between the two percentiles. The calculations are performed using the Excel *percentile* function. The way to summarize this data is, "The median VTW for delivery is 0.98 days early, while 90 percent of deliveries are between 8.09 days early to 5.51 days late." (Fig. 5.46)

The span metrics can be broken out by product line, sales region, or other such business unit. Figure 5.47 shows a graph with the span on VTW for product delivery broken out by delivery priority. The median for normal priority orders is one day early, while the median for emergency orders is 0.2 days early. If we only considered the measure of central tendency, it would seem that the

	A	B	C	D
1	VAR_TO_WANT			
2	-2.655		P5	-8.09
3	97.628		Median	-0.98
4	-1.618		P95	5.51
5	-2.965		Span	13.60
6	-4.209			
7	-8.999		P5	=percentile(a:a,0.05)
8	-3.996		Median	=percentile(a:a,0.50)
9	-1.321		P95	=percentile(a:a,0.95)
10	3.408		Span	=D4-D2
11	-0.982			

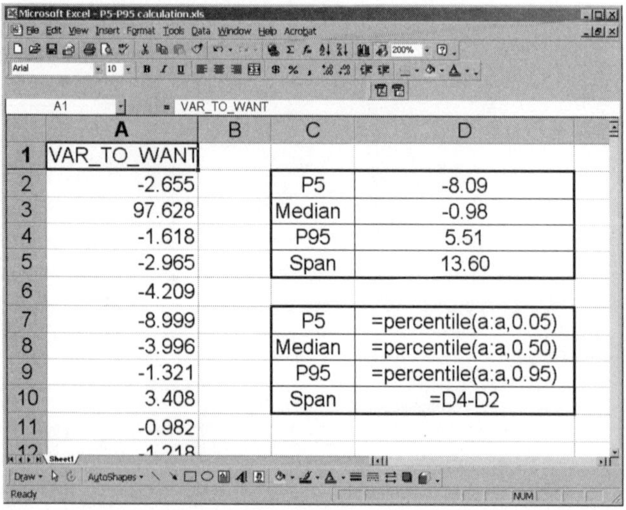

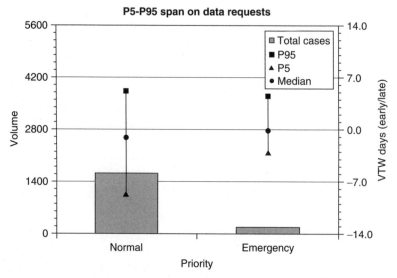

Figure 5.46 Calculating P5, P95, Median, and Span Using Excel

emergency orders are only handled at the last moment. When we investigated further we found, quite naturally, that emergency orders are quite short term and quite unlikely to be ordered with greater than about a week's lead time. The span on VTW for emergency priority product deliveries (7 days), however, as about half the width of normal priority orders (14 days) indicates

Figure 5.47 P5-P95 Span for Delivery to Customer Want Data (VTW)

that emergency orders are more consistently delivered close to the customer want date than normal priority orders.

The exact values for the percentiles defining the span will depend on the organization. When GE initiated the span metric, Bob Nardelli, then the CEO of GE Power Systems, stated that all customer touching processes should have a *P5-P95* span target for VTW of ten days. When GE corporate started receiving reports from other GE businesses using *P1-P99* or *P10-P90*, it took a corporate edict to define span as *P5-P95* across all business units. As projects were identified and executed, and business processes became more predictable, the metric was changed to *P1-P99* across the corporation.

5.17 Hazard Plots with Customer Want Dates

The hazard plot for the VTW data from the heavy equipment manufacturing project showed a time dependence not seen in the cycle time (Fig. 5.48). Orders were processed at a constant, slow rate when they were not due in the near future, then rate picked up as the customer want date approached. The shock to the project teams was that as the orders become late, the likelihood that they would be delivered dropped. After about 10 days past the customer want date, the rate was similar to that seen 10 days before the customer want date. Another way to express this is that the orders were processed more quickly as the due date approached, but given the choice

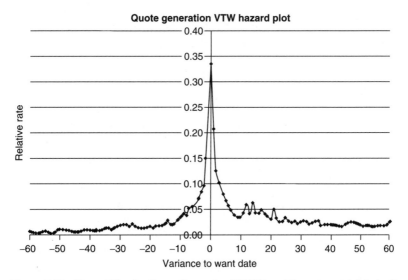

Figure 5.48 Hazard Plot for Quote Generation VTW Data (Nonparametric Method)

between doing something that was due immediately and something that was already 2 weeks late, the immediate order would be processed first.

5.18   Validate Your Assumptions

We conducted a lean Six Sigma project for a manufacturer of large equipment because irregular delivery was a constant complaint from the customers. When we were measuring the difference between expected and actual delivery times, we found that the differences were never more than about 14 working days. We went further into the data and traced the pedigree of the delivery dates in the various sales, production, shipping, and finance computer systems. Many of the final deliveries had been arranged by the customers. They made arrangements for their own shipping companies to pick up the orders and deliver them to their various job sites. The manufacturing business knew it was difficult to confirm the actual delivery times so they build in a *default value* for the delivery date. The default value was *Delivery date = Shipment date + 7 days* for domestic shipments, and *Delivery date = Shipment date + 14 days* for international shipments. The reality was that shipping large industrial equipment to international sites could take as long as a few months.

This problem was confounded by the performance metrics imposed on the sales managers that encouraged some sales managers to "ship" orders by placing the equipment in a shipping container, placing in their own parking lot, and contacting the customer's shipping company for pickup. Internal metrics for delivery-to-promise date looked fine as reported at the corporate level. The finance department would also find themselves in the embarrassing situation where they had generated a fairly large invoice for equipment that the customer had not yet received.

The problem had been known for years, but each organization along the value stream declared that a solution was out of their sphere of influence. When contacted, the customers declared that they did not want to know who was responsible, but were more interested in an accurate answer to queries about the status of the physical location of the equipment then they were about the cycle time for delivery. The solution was to offer the customer tracking service independent of the shipping company. A *global positioning system* (GPS) unit was added to the shipments, allowing the customer to track their orders while they were at sea, if necessary, and independent of the shipping company responsible.

Another portion of the data showed that about 3 to 5 percent of orders were shipped about 37,000 days late, while the sales reports failed to show any problems. The shipping database showed that the promised delivery date was January 0, 1900. We traced the shipments backwards through the information systems involved and found that the sales database had no promise date entered. The sales database would not include the order in the calculation of business metrics at the sales level, but as the promise date moved through the system, the blank became a zero and was reported as January 0, 1900 (date = 0).

Businesses are global; time zones, date formats, languages, calendars, currencies, and workdays are local. Call centers might be located in different continents than your customers. Delivering on a promise of "same day service" will require a measurement system that carefully tracks the time at the customer's site.

5.19   Gage R&R and the Data Audit

A Six Sigma project should include a phase where the reliability of the data is examined. In the manufacturing environment this can involve having two or more people, all measuring the same set of dimensions for a set of parts. Data gathered can be examining using *analysis of variance* (ANOVA). We have found that this narrow definition of Gage R&R is a bit confining and is not easily applied in all transactional projects.

The real purpose of this step is to force the project team to consider the data as reflecting not one, but two processes. The largest sources of error in the measurement process come from poorly defined and misunderstood operational definitions of the components of transactional processes, and assumptions made about data derived from historical systems that were never designed for the purposes of data collection. Without a good understanding of the measurement process as a separate process, the subsequent analysis can be deceiving or a waste of time and effort.

The project focus must always remain on reducing defects as defined by the customer, but be careful not to lose sight of the measurement process for the sake of the business process. During this phase of the project you must resist the temptation to start analyzing the data and concentrate on understanding it. The project consists of two layers of processes: the business process and the measurement process (Fig. 5.49).

Much of the portion of your lean Six Sigma project will be centering around cycle time, execution time, and delay time. It is natural to assume that the time/date stamps on data in a database will be valid, in the sense that the data was created at a particular point in time, but you will have a difficult time coming up ways of repeating the same transaction multiple times to obtain the data required for a full ANOVA, Gage R&R treatment.

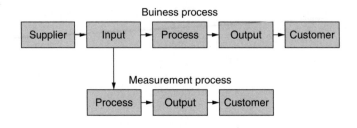

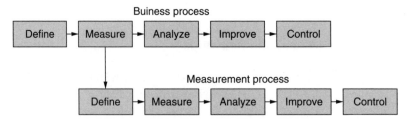

Figure 5.49   The Variation of the Measurement Process Overlays the Variation of the Business Process

5.20   Does Your Data Make Logical Sense?

During a project at a courier company, we followed the physical process and expected that the sign off data would be quite accurate and precise. It was usually measured without human intervention by barcode scanners. When we traced a number of shipments using the company's web tracking software, it gave us a detailed list of locations and time/date stamps that told a different story. The summary data for the shipment of an airline ticket from American Airlines in Texas to the author in Alberta is shown in Fig. 5.50.

From the customer viewpoint we surmised the following:

- Was picked up in Grand Prairie (located about halfway between Fort Worth and Dallas, close to the airport)
- Arrived at the Dallas/Fort Worth airport
- Left Dallas/Fort Worth airport (bound for Memphis or Grand Prairie) and then left Grand Prairie about 20 minutes later
- Perhaps traveled directly from Grand Prairie to a central sorting facility in Memphis
- Left Memphis twice to travel to Calgary
- Arrived in Calgary

```
Tracking Number : 494074493597
Reference Number :
Ship Date :
Delivered To :
Delivery Location : CALGARY CA
Delivery Date/Time :
Signed For By :
Service Type : Priority Letter

Scan Activity Date/Time Scan Exceptions
------------- ---------------- ---------------
Arrived at FedEx Ramp CALGARY CA             10/20/2001 07:27
Left FedEx Sort Facility MEMPHIS TN          10/20/2001 00:59
Left FedEx Sort Facility MEMPHIS TN          10/20/2001 00:52
Arrived at Sort Facility MEMPHIS TN          10/19/2001 23:47
Left FedEx Origin Location GRAND PRAIRIE TX  10/19/2001 23:38
Left FedEx Ramp DALLAS TX                    10/19/2001 23:12
Arrived at FedEx Ramp DALLAS TX              10/19/2001 21:29
Picked up by FedEx GRAND PRAIRIE TX          10/19/2001 15:23
```

Figure 5.50 Tracking Results for a Package Traveling from Grand Prairie, Texas to Calgary, Alberta

This is an example where a formal gage R&R was not conducted, but we still determined that the data may be unreliable. In this project, we determined a set of rules that allowed us to derive the cycle time data we required to characterize the process steps.

5.21 Capturing Rework

When a transaction is found to have an error, it is usual to send the offending transaction back to the previous step in the process for correction. Two things generally happen, the time spent correcting the error is logged as time spent processing the next transaction, and the time stamp for completion of the first transaction is overwritten when it is processed for the second time. Figure 5.51 shows the effect on cycle time measurement when an error is discovered and sent back in the process for correction. In this example, the data collection systems have been designed such that start and stop times are overwritten when passing through a process step for a second time. The effect in this case would be that the cycle time for all parts except one (Part D, upper lane) would have a recorded cycle time of three time units. It would appear that only part D in the upper lane was slower than the rest and took six time units while part C took only three.

Another project involved a technical help line that processed complex requests requiring research and further information from customers. Some customer service representatives would wait until the promise date before giving a partial

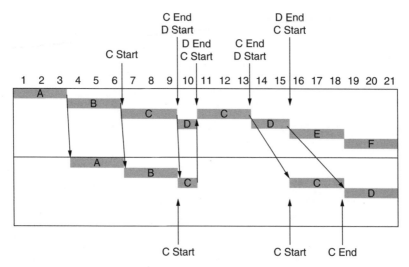

Figure 5.51   The Result of Rework Not Being Captured Correctly

response and requesting more information. Additional information was usually required, but the original request was signed off as occurring on time. The subsequent query was entered as a new query with its own promise date. The system was changed to track the second response as linked to the first while maintaining the original promise date. The performance metrics for service representatives were redefined to measure the variance to promise date for a complete response as measured by the customer via an e-mail response.

5.22 Bad Dates

We would not have dedicated a single section to one source of errors in lean Six Sigma projects if it had not come up as a major problem in every transactional project we have ever done. Accurate time/date information is the primary metric for the measurement of cycle time. The convention for recording dates using a format such as NN/NN/NN is ripe for a multitude of errors. Date conventions used by popular software programs are affected by operating system settings and can change depending on the user. Software filters may be present to reject an entry with a month greater than twelve, or a day of the month greater than 31, but these filters will not allow the software to correctly distinguish between 03/05/04 and 05/03/04. The default setting for the interpretation of NN/NN/NN can be altered using the regional settings on the control panel in Windows, so even the same entry on two different machines can be interpreted differently by data entry programs. The examples in this book are on a machine set with "English (US)" conventions.

The set of comma delimited data shown in Fig. 5.52 came from four separate financial processing systems all feeding a single database. The dates in the file were interpreted by the database during the import and made it difficult to sort out the sources of bad data. We had to examine the raw data files from each of the four financial subsystems before we could determine the drivers of unreliable data.

Different system settings on the individual machines resulted in different date and currency formats during the export process. Individual operators would also change system settings on single systems. We had to work with the individual computer systems, standardize system settings, and make sure they were retained after a shutdown and reboot of the financial subsystems.

The problem of merging data from different sources can be masked and compounded further by properties of the software being used to merge the data. It is very common to use MS Access or Excel to extract and manipulate data. When data is being entered or imported into a particular cell, the application will interpret the entry as a date if it is a recognized date format. If an entry is recognized as a valid entry, then the value is converted to the number of days where day 1 is equal to January 1, 1900 (Macintosh Excel, day 1 = Jan 2, 1904). The stored value will now be displayed using the date format chosen for the cell, if applicable. For example, in Fig. 5.53, the first entry is recognized as a valid date, converted to the internal value of 38,709, and displayed using the imposed format of MMM DD YYYY.

Entry number 2 shows that Excel cannot interpret "dec" if it is in the first position in the string. Since Excel cannot interpret the entry as a date, it accepts the entry as a text field and displays it as text regardless of any imposed formatting. The deception is that the displayed record appears the same as entry number 1, a correctly interpreted record. The only clue we might have is that entry number 1 shows the capitalized "Dec" for the month, while entry number 2 has been passed through without change. Be aware that even numerical data may be interpreted as text. Check everything.

```
$218,975,"SW",,,"211",11/17/03 3:28:53 P"
$3242.00,"SW",,,"212",,,
$20,742.01,"Central",,,"212","17/11/2003 3:30:52 PM"
18451,"SW",,,"213","03/11/17 3:31:14 PM"
653678,"Asia",,,"214","17/11/2003 15:29:00"
```

Figure 5.52   Different Data Formats for Raw Data From Four Different Financial Reporting Systems (System 211, 212, 213, and 214)

Entry Number	Entered Data	Interpreted Value	Displayed using MMM DD YYYY
1	23 dec 2005	38709	Dec 23 2005
2	dec 23 2005	dec 23 2005	dec 23 2005
3	2005/12/23	38709	Dec 23 2005
4	2005/23/12	2005/23/12	2005/23/12
5	23/12/2005	23/12/2005	23/12/2005
6	12/23/2005	38709	Dec 23 2005
7	2005/12/23	38709	Dec 23 2005
8	2005/23/12	2005/23/12	2005/23/12
9	0	0	Jan 00 1900
10			Jan 00 1900

Figure 5.53   Software Interprets Some, But Not All Date Information

Another problem is the common use of NN/NN/NN date formats. The year is commonly placed last, but there is little universal agreement whether to start with the day or the month. The common convention in use shows some global variation, but is unreliable given that workers have widely different backgrounds. Maximum value filters will allow some incorrect data to be entered and misinterpreted while other entries will be retained as text fields. Subsequent calculations will return invalid values for some cycle times even though the displayed date values appear valid (compare entries 1 and 2 of Fig. 5.53).

We had one occasion where all maintenance records used the Muslim calendar, while other records used a variety of international conventions of the Julian calendar. Consequently, cycle time calculations showed some maintenance problems that took management about 580 years to acknowledge.

Another case is that software can interpret both "0" and blank as "Jan 00 1900" (entries "9" and "10"). When the date for a particular process step is unknown, it can create problems for cycle time calculations if the start date is not entered or is entered as a zero. The time portion of a date/time stamp has similar problems associated with computer programs interpreting input. A date/time entry can be entered using any valid date format followed by a number of different time formats. The time can be either on a 12 hour format with an implied or explicit *a.m.* or *p.m.*, or a 24 hour format. When an *a.m./p.m.* field is not entered, the programs may have unpredictable interpretations about the time of day.

Even when all software interpretations of times are understood, great care should still be exercised about the human interpretation of the meaning of 12:00 a.m. and 12:00 p.m.

5.22.1 Checking for Date Errors

When checking time/date stamps, it is helpful to pull all the raw data from a database in sequential order and calculate the elapsed time between each data point and the next. All differences should be small and slightly positive. Incorrect date formats show up as large positive or negative jumps. Figure 5.54 shows run charts of transactional data for the scanning and electronic filing of documents as part of a medical claims process. The cycle time examined here

Cycle Time from Claim to Receipt of Documents

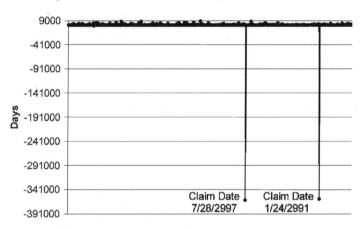

Cycle Time from Claim to Receipt of Documents

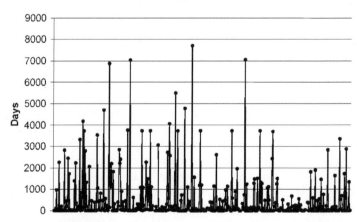

Figure 5.54   A Run Chart of Cycle Time Data Shows Aberrant Data Points. Typographical Errors Were Corrected Resulting in a More Reasonable Picture of Cycle Time

was the elapsed time in hours between the receipt of the document and when the final image is indexed and uploaded to a server. The elapsed time was calculated as the difference between the data fields for *indexdatetime* and *receivedatetime* on the corporate data server.

The run chart of the time ordered data in the upper half of Fig. 5.54 allowed us to rapidly identify the two aberrant data points listed in the figure. The source of the problem was a data entry system that accepted the operator entered the date using YY/MM/DD format. During the upload to the server, the date was translated into the YYYY-MM-DD HH:MM.

The emphasis of the project was not one of identifying the drivers of bad data, but one of identifying the bottlenecks in the process. The Measure phase consisted of a step to screen the data before characterizing and analyzing the "clean" dataset.

5.22.2 Quantifying Date Errors

When you are benchmarking your measurement system for date/time stamp information, you will undoubtedly run across entries with incorrect dates. Quantify the magnitude of the error by using a *net present value* (NPV) calculation with the transaction.

$$\text{Interest} = \text{principal} \times (e^{\text{time} \times \text{rate}} - 1) \qquad (5.15)$$

For example, if a $100 payment was due on March 12, 2005 (03/12/05) but had been incorrectly entered as December 3, 2005 (12/03/05) and the interest rate was 10 percent p.a., the magnitude of the error would be the lost interest for 266 days ($7.56). Calculate the total variation in dollars, both positive and negative (Fig. 5.55). For details about the calculation of span, see Section 5.16.

The example shown in Fig. 5.55 was based on the data analysis of the project directed toward reducing the amount of time the sales team spent doing administrative tasks rather than spending time with customers (Section 5.12.4). We found that sales representatives were spending anywhere from 2 to 12 hours per week tracking and checking sales made to their sales regions. When the sales representatives requested weekly data extracts from the financial tracking system, they would find errors of omission as a result of the date cutoffs used during the data extraction. Data was often reentered by the sales representatives when they did not find transactions. The financial group would assemble month end statements by extracting all sales records

Entry Number	Entered Date	Real Date	Error(days)	Transaction Value	Interest
1	01/02/05	02-Jan-05	0	$20.00	$0.00
2	05/15/01	15-Jan-05	-1341	$100.00	-$36.74
3	01/03/05	01-Feb-05	-29	$340.00	-$2.70
4	05/02/01	02-Feb-05	-1372	$10.00	-$3.76
5	02/05/15	15-Feb-05	3642	$5.00	$4.99

Transactions	$475.00
P5 (error)	-$30.14
P95 (error)	$3.99
Span (error)	$34.13

Figure 5.55   The Date Errors Result in an Error in Timing for Transactions. The Magnitude of the Error is the Interest Lost or Gained. The Above Data was Calculated with Simple Interest at 10% P.A

created since the last time they were extracted, regardless of the date in the system. It was typical that the finance group would take about 3 to 4 days at the end of each month reconciling sales figures before reporting them at the corporate level.

The solution was to reconfigure the sales tracking software to use the DD-MMM-YYYY format for dates and to use the server date as a default value. The benefits on the project were:

- An accurate value for sales, as quantified using the method shown in Fig. 5.55.
- About one extra day of productivity per week for the sales representatives.
- About two days more productivity for the finance team.
- The cycle time for preparation of month end sales statements went from a couple of days to a few hours.

The net financial gain for the project was over $100 million in extra sales revenue, all as a result of correcting date formats.

5.23   Quantifying Overwritten Data

While we were conducting the data collection plan on a financial approval process, we encountered a characteristic of the tracking system illustrated in Fig. 5.52. Credit check and escalation approval dates were overwritten when data was uploaded to the *customer relationship management* (CRM) system. In order to use the existing systems for gathering data, we would extract sign off times from the raw data before they were uploaded to the

CRM. Loan applications with more than one unique sign off time for the same step in the approval process were flagged as rework.

The impact of the overwritten data was quantified as undocumented rework during the Measure phase of the project. The elimination of the now visible rework was reported as the benefit of the Improve phase.

5.24 The Measure Checklist

This chapter has spent a lot of time discussing the definitions of metrics, data requirements, and the focus on transactions at the customer level. It is a common credo that you get what you measure. It is extremely important to think very carefully about the metrics you are measuring. These definitions will persist throughout the project. When the metrics you are using become part of personal performance appraisals then they can become altered. When we started measuring the VTW on orders, we noticed that the size of orders shifted to smaller items. The CEO and CFO became very concerned that the product mix had changed. The sales staff had simply divided large orders into lots of smaller ones to make their metrics on VTW for delivery look better. The customers had 95 percent of their order when they requested it, but the last 5 percent of vital parts would become much later than when the order was consolidated. Net sales seemed to increase at the same time.

Your Measure phase will consist of three macroscopic parts:

5.24.1 Select the CTQs

In theory, if you truly understand the customers' view of your business, it should be easy to define the problem. In practice this step is one of the most difficult and critical for project success. There are many stakeholders that will resist shining a spotlight on an area of their responsibility. Set aside any paradigms you may have and really focus on the customer. Accept that a well-scoped project will include only a portion of the total customer viewpoint. The largest barriers will come from your own perceptions of what is needed, possible, obvious, or affordable.

At the beginning of this step:

- What are your perceptions? (For example, cost is not the same as value, and speed is not the same as satisfying an expectation)
- Is this a new problem? What has been attempted in the past?
- How important is this to my customers?

At the end of this step:

- Has your problem definition addressed any discrepancy between current internal and external views (For example, airlines do not "lose" baggage, they "misdirect" it. Days versus workdays)
- Which physical business processes are targeted in your project?
- If the project appears to be too large, it should be split up into coordinated parallel projects.
- If you are separating the project into smaller pieces, can different CTQs be handled in separate projects?

Points to remember:

- Do not build your solution into the definition of your problem. (For example, "The sales team lacks a web-based tool to...")
- Do not try to redefine or develop a metric as a way of defining the problem.
- Always focus at the transactional level when articulating the problem. (For example, You may have to measure VTW by line item rather than order number)
- Validate your assumptions on operational definitions. (For example "on time departure")

5.24.2 Define Performance Standards

It is likely that high level metrics used to measure how the business is performing will be inadequate for your project. The numbers reported at the corporate level and circulated quarterly or yearly give an aggregate view. Most of the individual transactional detail is lost in the consolidation. The existing metrics can mask problems at the transactional level. Customers make decisions at a transactional level and the process is influenced at the transactional level. You must gather the data on potential factors (Xs) at the customer level to be able to do any analysis.

At the beginning of this step:

- Do the existing internal metrics measure variation as well as the average?
- How are the metrics calculated in detail?
- Does the metric calculation contribute to my problem? (For example, Are sales people compensated based on net sales or margin, or both?)
- Do you have a defect reduction target from the customer or from competitive benchmarking?

At the end of this step:

- Is your measurement continuous? (For example, strive for date and time information for timestamps)
- Will the measurement retain the customer view at a transactional level? (For example, use time between customer arrivals instead of customers per hour)
- If the measurement were visible real-time, would it drive activity? (For example, display board with average time to serve)
- Can the measurement be manipulated? (For example, capture cycle time including task reassignments and partial shipments of orders)
- Does the metric measure the variation in the process? (For example, use *P5-P95* span on VTW)

Points to remember:

- This step is all about how the customers measure each transaction.
- Whenever you capture data, attempt to capture the customer expectation for the transaction at the same time.
- Historical data collection systems were not designed for your project and may not measure what you require.
- Behavior may change simply because you are emphasizing a particular measurement. Continue to measure using existing systems to detect if you are only moving the problem elsewhere.

5.24.3 Validate the Measurement System

It is common that the existing measurement system can give unreliable or misleading data. This step is usually underrated during the project planning, but is one of the most important for subsequent analysis. When you present your team's conclusions at the end of the Analyze phase, your results may run contrary to company folklore. Your project will usually be attacked for the integrity of the original data rather than the depth or sophistication of the analysis. It is important that you satisfy the stakeholders that your data is reliable and represents a good cross section of sales regions, seasons, product mix, customer type, business units, and so on. Get them to agree to the data collection plan and subsequent analysis before proceeding to the Analyze phase.

At the beginning of this step:

- Where and how are the data recorded? Try to drive back to the primary sources.
- Are all records collected in the same manner? Is there a possibility that two different business units record similar data, but for their own purposes?

- What is the maximum error on the data that customers will accept? (Bank transactions arriving after 2:00 p.m. are credited on the following business day)

At the end of this step:

- Get the stakeholders to agree that your data collection plan has resulted in a rich, unbiased set of data that represents a cross section of the transactions between the business and the customers.

Points to Remember:

Lean Six Sigma has statistical tools to verify the integrity of the data itself. Use either the attribute (discrete data) or variable (continuous data) gage R&R. If this is not possible, go back to your data with your operational definitions and ask the following questions:

- What does your data represent?
- Who has access to it and can alter it?
- Are you looking at the data at a sufficiently granular level?
- How are signoff date/time stamps treated for rework? Are they overwritten?
- What are the assumptions with blank or absent data?
- What are the effects of customers or facilities in different time zones?
- Do different parts of the business have different days and hours of operation?
- Do you have consistent units of measure? (boxes, cartons, crates, tones, tons, kilograms, pounds, US$, CAN$, Euro)
- Have you traced the data beginning from the primary source to the final data system and assessed the assumptions made at each transfer?
- Make sure you understand date/time conventions for each source of data.
- Keep asking questions about the source of the data; validate your assumptions even though you think they are obvious to all users, team members, and customers.
- The time you spend on this step will have enormous pay backs during the Analyze phase. The analysis will proceed faster, will be more reliable, and be accepted by the stakeholder more readily.
- Even if your gage R&R is acceptable, include a periodic data audit as part of your control plan.

6

Analyze

6.1 Customer Demand and Business Capacity

One of the GE businesses conducted a number of surveys of customers to find how they defined quality. Initially, quality was defined by such things as,

- Product or service technical performance
- On time, accurate, and complete deliverables
- Customer responsiveness and communication
- Marketplace competitiveness

These issues are important, but they were necessarily vague to cover the great variety of products and services offered by a large and diverse company. When these metrics were applied at the individual business unit level, the definitions would be interpreted differently for different product or service offerings, different customer segments, and such.

A problem that arises in lean Six Sigma is that as businesses become large, project teams become more and more removed from the real customer. Recognizing who speaks for the customer in business-to-business transactions can become difficult. Is it the engineering, production, sourcing, finance, or executive groups?

Another problem that arises is concentrating on business improvement efforts is only one aspect of the relationship between the customer and the business. Decreasing the causes of waste, or *muda* in the lean lexicon, will decrease costs and cycle time, and the resultant agile organization can react to customer demands more quickly with less waste. Cycle time, however, is not the primary metric for a business process, and cycle time reduction is not the primary metric for process improvement. The total story in satisfying and managing customer needs requires measuring, analyzing, and understanding the complex relationships between:

- Customers' demands
- Customers' expectations
- Production capacity
- Employee performance metrics
- Product or service quality
- Sales revenue and growth targets
- Customer satisfaction

The business literature is filled with examples of companies that applied changes to their business processes to address one or two of these points, yet still failed in the long run.

Customers consider a large number of factors when making a decision to buy from your company. This depends on a complex balance between your service's price, value, service level, economics, risk, need, features, reputation, and similar assessments of competing service providers. Most sales staff have been encouraged to agree to customer demands to ensure a sale. This behavior may be encouraged by incentives or the sales department's genuine confidence that the service department can deliver on the agreement.

Once a decision has been made and sales has made a promise of performance, the customer goes away with an expectation. They will return at some time in the future to complete the transaction. If you fail to keep those promises, the customer eventually ceases to deal with your company. The downturn in sales will be noticed by the sales executive who instructs the sales staff to make more promises to the customers, and the cycle continues. As more customers become dissatisfied, you lose market share to your competitors (Fig. 6.1).

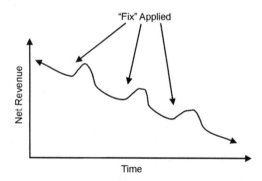

Figure 6.1   The Slow Death of a Company

The cause of this problem is usually a misconception that the customer must have something faster than your company can usually provide, so the sales people will make an unrealistic promise to the customer and hope that the order can be expedited. A company that promises less than it can deliver is less common than this situation, but still damaging. When a company has expended resources to make a shipment go out early, those are resources that should have been placed on a different shipment that was already going out late.

Lantech is a company that manufactures machinery for stretch wrapping large pallet loads. As described in *Lean Thinking*, in the early 1990s, the company made enormous leaps in restructuring based on lean manufacturing principles.[*] The product delivery lead time had been 4 to 20 weeks in the past. They describe one case where the new, faster production lead times had not met customer expectations.

"In one notable case, Lantech made and delivered a machine within one week of the order, as promised, to find the customer quite upset: 'You've sent us our machine before we've given any thought to how to use it. We thought we were placing an order simply to guarantee ourselves a place in the production line, that we would have to respecify the options, and that you'd deliver last as usual. Now, you've gone and made it already!'"

This situation and others like it illustrate the complexities of the relationship between the customer and the business. Time lags between the elements of the extended relationship between the customer and the business can produce positive or negative feedback loops that can reinforce or dampen attempts to change the overall process. In general, these phenomena were studied by Jay Forrester in the 1950s under the discipline of systems thinking as a field of systems dynamics. The business application of systems thinking is well reviewed by Peter Senge.[†] He describes the scenario resulting in the slow death of a company in Fig. 6.1 as "fixes that backfire." It consists of a quick fix resulting in a positive effect in the short run, but with a larger, negative effect in the long term. Another example of "fixes that backfire," is illustrated in a cycle of layoffs and falling net income in Fig. 6.2.

Let us start at the point where the decision was made to apply a short-term fix to the problem of gradually falling net income. The immediate result of workforce layoffs was that costs decreased. The capacity of the business to

---

[*]James P. Womack and Daniel T. Jones, *Lean Thinking*, Free Press, NY, 2003.
[†]Peter Senge, *The Fifth Discipline*, Doubleday/Currency, NY, 1990.

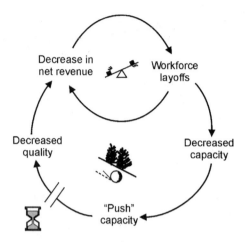

Figure 6.2   "Fixes that Backfire" Scenario From Senge, et al.

provide service also decreased, but management hoped that they could push the remaining capacity or restructure the business process such that production levels could be maintained with the smaller workforce. There was a time delay, but as the workers were pushed more and more, the quality of the product fell. Customers became dissatisfied and bought less product, net revenues began to fall again and management initiated another round of layoffs. The effect of repeated quick fixes followed by longer term detrimental effects produced the cyclical downturn of net income as shown in Fig. 6.1.

The symbols in the diagram show the time delay (hourglass) in the detrimental effect and the interaction of the two loops. The top loop is the balancing loop (see-saw) where the fix is applied to the problem, while the bottom loop is the reinforcing loop (rolling snowball) that generates the detrimental effect. The relative sizes of the effects of the top and bottom loops are responsible for the long-term effect of the layoff policy. The time lag is responsible for the "backfire" following the fix. The system can also be moved in reverse. If the quality of service increases, revenues will increase, which can be invested in extra capacity.

Senge goes on the explain how other common business problems can be modeled with different configurations of balancing and reinforcing loops with different time delays on each. These scenarios, called *archetypes*, include commonly encountered systems problems such as "shifting the burden," "limits to growth," "escalation," and "drifting goals." The archetypes can be combined to give an overall picture of a more complex system.

6.2   Accidental Adversaries at Company X

A common scenario can be used to illustrate how different aspects of an organization can contribute to degrade the overall quality of service. This example is based on a real situation that existed for many years. It resisted an intense effort at improving the situation with a large number of sub-optimal Six Sigma improvement efforts.

The systemic problem was an eroding reputation of the organization to deliver service to the customers. The service was extremely individualized and the value of the individual service contracts differed by as much as four orders of magnitude. A survey was conducted among the executive of the customer base. Typical responses were, "Your company is great in an emergency, but day-to-day service is very poor," and "We'd be happy if you could just deliver what you said you would deliver." The system of "accidental adversaries" in the company is shown in Fig. 6.3.

Sales targets were given to the sales staff, tracked weekly, and reported monthly. Sales compensation levels were measured quarterly and awarded yearly. At the beginning of the month, the members of the sales department would look at their sales targets and plan their sales strategy. When contacted, customers would ask for estimates of cycle time and price. The sales department would base their negotiating position on their experience with the "awesome" service department who had developed a reputation within the company for stretching to meet customer requirements. The sales

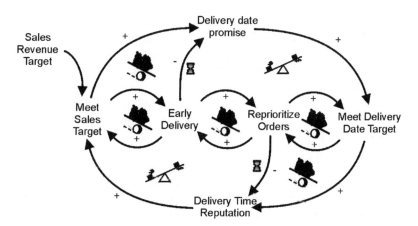

Figure 6.3   Accidental Adversaries Between Sales and Service

department was convinced that getting the customer to commit to the order was going to depend on the performance of the service department. The sales department's perception was that it took too long to get a good estimate of price and cycle time from the overworked service department, but their experience was that they usually produced conservative estimates and delivered underprice and early. They would agree to the customer's needs and promise a delivery date.

The order would arrive at the service center, and realizing they are already swamped with backlogged work, they would inform sales that the promise date could not be met. The reaction would be that favors would be called in, management would intervene and want special treatment for this customer, the order would be broken up into partial shipments orders, price concessions were made, or the order would be expedited. Service would be held to the published cycle time, they would delay other less critical orders, and put extra people on this particular order. The order would usually come in a few days early while everyone in service and sales would be hailed as heroes. The customer would be happy and the delivery would enhance the company's reputation for customer centric focus. Sales would make their revenue target. The overall philosophy was that the company would fix the system—*one customer at a time*. The reputation of the service department to handle *special* cases began to become the norm and was expected by the sales team.

When the service department reprioritized the orders within the system, the likelihood of meeting the due dates of all the other orders would now be in jeopardy. Price and cycle negotiation limits become wider as the sales department got more experience in handling special orders. As the number of special orders increased, the capacity of the service department to reprioritize diminished to the point where orders handled with *normal* priority were so late that the reputation of the company began to suffer. Sales begin to suffer and more aggressive targets are given to the sales department. An adversarial environment between sales and service grew over the years to the point where the senior managers would not attend common meetings unless ordered by the executive. Each side was convinced that the problem was caused by the other.

The company culture had slipped into a philosophy that sounded good—the company was customer focused and responsive. The business had devolved from one that offered the optimum mix of cost-efficiency with the flexibility of product and service offerings. It had changed from a batch flow-type company with efficient, loosely connected subprocesses to one with disconnected processes required for a low volume, highly customized offering, but without enjoying the financial benefits of producing a one-of-a-kind product (Fig. 6.4).

The negative factors in Fig. 6.3 had shifted the problem from managing a single transaction to mismanaging a large number.

This chronic problem had resisted a number of suboptimal solutions. With the support of a senior and corporate executive, we took a more global approach and gathered data on the cause and effect relationships between all of the different elements. The solution required changes in sales compensation and updated capacity planning to restore the business to one with the best combination of responsiveness with cost-effectiveness.

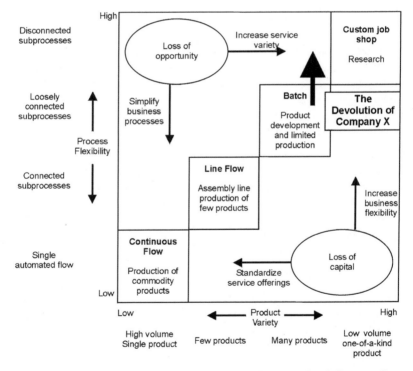

Figure 6.4 Company *X* Had Slowly Moved From a Balanced, Batch Company Into the "Loss of Opportunity" Area

As the improvement was implemented, we designed reporting metrics that could identify when

- The service department was backlogged and required extra capacity.
- The sales department was making unrealistic promises to make a sale.
- The service and sourcing departments were basing estimates on unrealistic values for cycle times.
- Too many customers were asking for special treatment than could be handled by the system.
- Orders were being reprioritized incorrectly once they were in the system.
- Booked sales were masquerading as customer needs.
- Customer want dates were renegotiated after the order was placed.

Each one of these changes would have a short-term benefit on a department or customer, but would have a long-term detrimental effect on the entire customer base.

6.3 Lessons from System Dynamics

The dynamic behavior of simplistic systems like the "accidental adversaries" and the "fix that backfires" can be stable, oscillatory, or chaotic depending on the parameters of the feedback loops. Constructing models for more complex dynamic systems and prediction of their behavior is beyond the scope of this book. We can, however, make some general notes about the properties of the dynamic systems you will encounter.

- The overt behavior of dynamic systems tends to lead to oscillations.
- Each business system will have its own set of characteristic frequencies.
- One or two feedback loops dominate the dynamic behavior of the system.
- If the frequency of changes in the input signal (for example, customer demand) approaches the frequency of response (for example, supply) then the system resonates: oscillations build up over time.

The lessons relevant to lean Six Sigma are:

- Business processes are not always linear.
- Problems can be caused by the interaction of business processes.
- Chronic problems must be considered holistically.
- Long-term fixes require high visibility of the impact of short-term decisions.
- Quick reaction to problems in *variance to want* (VTW) may lead to problems with capacity and long-term impact on customer.

- Failing to see the big picture can result in a problem being "fixed" while inadvertently creating a new problem elsewhere in the organization.

6.4 Analyzing the Entire Problem

The previous examples have shown that concentrating on a single aspect of a business can lead to eventual failure owing to the complex interrelationships between the multiple elements of any business process. This often happens when traditional Six Sigma projects are executed in only one area of a business. It is termed suboptimization in the systems dynamics literature and "throwing the problem over the wall" in the practical world.

This extended view of the business is also important because a single transaction will touch many areas of the business. If we neglect components such as warranty or aftermarket service, we can define a single transaction into three steps:

1. *Inquiry to Order* (ITO)
   - Customer and sales make contact.
   - Customer expresses wants (date, price, quality).
   - Sales check with service for availability and gives customer an estimate on deliverables (requires knowledge of capacity).
   - Customer accepts or rejects offer of service (could involve a quote to the customer).
2. Customer makes the order (or converts the quote to an order)
   - This step signifies that the customer has accepted the promises of performance made by the business. This establishes the customer wants and sets the goals for performance for the next phase. The customer wants may not have all been satisfied during the ITO step, but at this stage the wants now become expectations.
3. *Order to Remittance* (OTR)
   - Schedule and assemble the order according to the promises of performance.
   - Customer accepts delivery.
   - Invoice sent to customer.
   - Customer remits payment.

All customer expectations are defined at the moment of placing the order, with few exceptions. Any customer wants that are missing during the OTR step are usually considered as errors of the ITO step. This process is shown as a linear one, but at least one feedback loop is present. This feedback loop has potential to create the long term detrimental effects seen in the "fix that backfires" or "accidental adversaries."

6.5   Establish Process Capability (P5-P95 Span)

In the early stages of the Analyze phase of a traditional Six Sigma project, you will be establishing the process capability. This calculation involves combining the variation in the process with the *upper specification limit* (USL) and *lower specification limit* (LSL) from the customer. These two aspects are combined to produce the *defects per million opportunities* (DPMO), or process capability—the capability of the process to produce product within the customer specifications. This measurement can be expressed in a few different ways, DPMO, Sigma level, Cp, or Cpk. The exact manner of reporting is less important than the measure of impact on the customer. The calculation can be "run backwards" if only attribute data is available, but this calculation is much less diagnostic. The relationship of the process variation, USL and LSL, and defect levels are shown in Fig. 6.5.

In the highly repeatable processes usually encountered in high volume manufacturing, the specifications for the output of a single business process do not change with each piece. The measurements of many parts can be summarized before comparing the process variation with a single pair of specification limits. As the business moves more towards specialized products, different classifications of parts can be consolidated to provide a large enough sample within a product offering to summarize the variation within the group. If we were to extend this concept into transactional business processes, an example of this segmentation might be to divide loan processing times by "size bucket" (small, medium, large) and customer segment (commercial, residential, personal). When there are too many segments, however, it becomes difficult to combine the multiple measurements for summary or analysis.

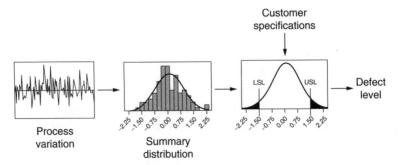

Figure 6.5   The Relationship Between Specification Limits, Defects, and Process Variation for Process Capability in Manufacturing

Our focus on measuring VTW requires each transaction to "travel" along with its individual specification limit, the customer want date, quote price, or other customer expectation. The summary of VTW for the process will usually not follow any well-known distribution, so the calculation of defect level shown in Fig. 6.5 would be extremely difficult. We will only assume that the distribution is a continuous function and report the 5th and 95th percentiles. This metric is tracked and reported.

It is possible to impose a company wide limit in span of VTW such as, "The P5-P95 span on VTW for deliveries should be 10 days." When the span on VTW is reported back to the customers, it is a piece of information that becomes part of the individual transactions at the customer level. It is a metric that means more to the customer than a DPMO or sigma level. The customer will decide whether to deal with your company and enter into a transaction given this information.

Imagine that you are making a decision on a business trip that passes through O'Hare Airport in Chicago (ORD). Which metric would you prefer to see before making your choice of connecting flights?

- 73 percent of our flights are on time.
- 90 percent of our flights leave between 4 minutes early (*P5*) and 13 minutes late (*P95*).

The calculation of the span of VTW becomes our measurement of process capability for lean Six Sigma projects as shown in Fig. 6.6.

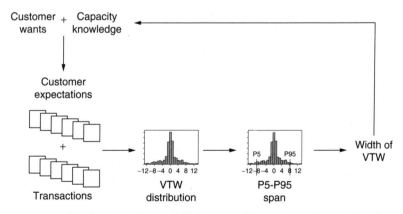

Figure 6.6 The Relationship Between Customer Expectations, Process Variation, and Span on VTW in Transactional Processes

6.6 Examples of P5-P95 Span on VTW

6.6.1 Variance to Schedule

We chose all flights scheduled to arrive at Calgary International Airport (YYC) before 7:00 p.m. on one day in February. The expected arrival times were those posted by the airlines days before and did not reflect any delays anticipated by leaving later than the published schedule. Severe weather in eastern North America was affecting incoming flights. Even though the median arrival time was three minutes early, the span was nearly one hour (see Fig. 6.7).

6.6.2 Variance to Quote Price

A small electronics manufacturing plant sells a selection of electronic components for use in larger hardware installations. They would sell components through reciprocal arrangements with other manufacturing companies and directly to customers for existing installations. Prices were published in the product catalogues on the company's website. They had several different discounting agreements where the discount would depend on the customer, the age of the product, the foreign currency exchange rate, sales volume, and length of time in dealing with the customer (preferred status).

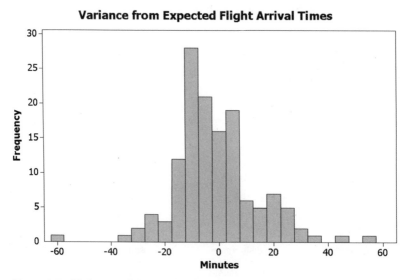

Figure 6.7 Variance to Schedule for Aircraft Arrival Times

The principal problems in the business were the accurate prediction of revenue based on orders and the inconsistent application of pricing policy. We gathered all sales for a calendar year and applied the published prices and pricing policies. We calculated the variation between the calculated discount price and the line item price from the invoices. We also expressed the variance as a percentage of potential revenue. The baseline values for span are shown in Fig. 6.8.

The inconsistent application of documented pricing strategy by individual sales representatives resulted in a *P5-P95* span of $69,777 per line item, or 76.5 percent per line item. The median variance from quoted prices was very near zero. The VP of sales was aware of this situation, but since the line item sales were consolidated for weekly reporting and margin review meetings, the variation was hidden from the CEO and CFO.

6.6.3 Variance to Want Date

A manufacturing of large industrial equipment has a customer contact center where a customer can make a query by e-mail, fax, or telephone. The agents gather information and pass it onto technical experts who research the information and have it reviewed before replying to the customers at a later date. The queries may entail upgrading equipment, maintenance, or operating recommendations (Fig. 6.9).

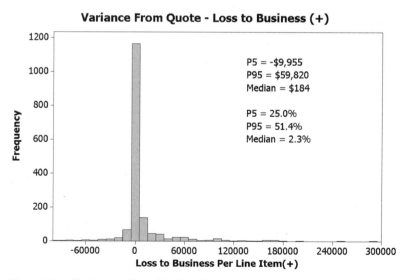

Figure 6.8   Variance to Quote Per Line Item

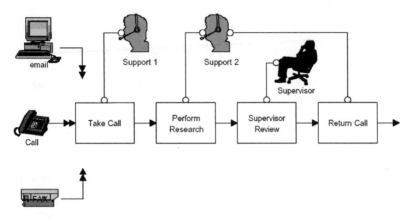

Figure 6.9   Customer Contact Center

The customer want date/time is recorded for each incoming contact. When the customer receives the information, the date/time is recorded again. We calculated the variance from the customer want date/time. Following the usual convention, early responses are negative while late responses are positive (Fig. 6.10).

The histogram showed the distribution was very peaked. Although the median was 0.97 days early, there were a number of requests that were on

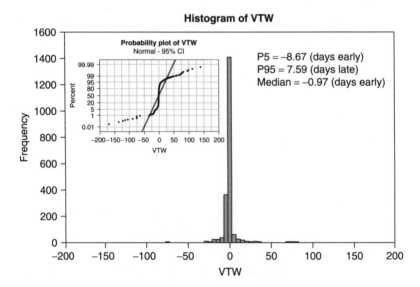

Figure 6.10   Variance to Want Date/Time for a Contact Center

the order of 100 to 200 days early or late. The *P5-P95* span was quite good at 16 days. When processes begin to show less variation, then span can be calculated using *P1* and *P99* to capture the remaining variation in the tails of the distribution.

6.7 Elements of VTW-Customer Wants

The multiple relationships between different elements of VTW require us to do an analysis of the following:

- Analyzing customer needs
- Analyzing production capacity
- Analyzing VTW
- Anticipating changes in customer needs and production capacity

We will be examining customer demand from a number of different perspectives. The raw data will be in the form of customer arrival times. This should be in date/time format and be as precise as you can get it. Even though we will be analyzing this data in hourly, daily, or weekly intervals, it should be recorded at least to the nearest minute and ideally to the nearest second.

A reasonable question is whether there are changes in the arrival rates of your customers. In statistical terms this is called testing for a homogeneous process. Customer arrival rates usually change from day-to-day during the week, or from hour-to-hour during the day. Longer-term variations are called *seasonal*. The data for the arrivals of customers at an imaginary bank during one day are shown in Fig. 6.11.

We produced the data by generating a random set of 500 arrival times, uniformly distributed between 8:00 a.m. and 4:00 p.m. The subtotals of the number of customers arriving each hour show that the arrival rate changes during the day. The arrivals seem to peak between 9:00 to 10:00 a.m. and after lunch (Fig. 6.12).

We suspect that the daily variation might not be random, and will use the Chi-square test to see if the differences are significant or not. The Chi-square statistic is calculated using the formula:

$$\chi^2 = \sum_{i=1}^{n} \frac{(\text{freq}_{i,\text{observed}} - \text{freq}_{i,\text{expected}})^2}{(\text{freq}_{i,\text{expected}})} \quad n - \text{the number of groups} \quad (6.1)$$

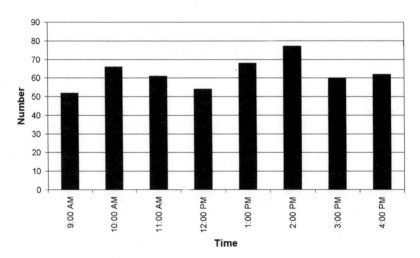

Figure 6.11   Customer Arrival Times

Customer	Arrival Time (HH:MM:SS AM)	Sorted Arrival Time (HH:MM:SS)	Interarrival Time (min)	Rate (Cust per min)
1	8:21:54 AM	8:02:48 AM		
2	10:52:08 AM	8:04:45 AM	1.949	0.513
3	1:01:40 PM	8:04:47 AM	0.035	28.566
4	3:43:59 PM	8:06:03 AM	1.264	0.791
5	12:33:00 PM	8:06:34 AM	0.530	1.886
6	2:02:02 PM	8:06:40 AM	0.089	11.188
7	8:59:22 AM	8:08:31 AM	1.853	0.540
8	1:47:53 PM	8:10:38 AM	2.109	0.474
9	8:06:34 AM	8:11:15 AM	0.629	1.590
10	9:14:37 AM	8:13:33 AM	2.292	0.436
11	11:26:47 AM	8:14:05 AM	0.541	1.847

Customers Per Hour

Figure 6.12   Numbers of Customers Arriving Throughout the Day

This can be calculated in Excel by totaling the counts in each bin using the histogram tool and calculating the values for the Chi-square test in the spreadsheet (Fig. 6.13).

If the number of customers arriving each hour was perfectly random, we would expect the same number of customers to arrive during each time interval. We had 500 customers, so we would expect the frequency to be 62.5 customers per hour throughout the day. These values are shown in column J. The calculation of the components of the Chi-square statistic is shown in column K, while the total Chi-square value, 7.104, is in cell K10.

All that remains is to use an Excel function to calculate the p-value. The formula "=CHIDIST(K10,7)" is entered in cell K11 to calculate the p-value for a Chi-square value of 7.104 with seven degrees of freedom. The number of degrees of freedom for this Chi-square table is n-1, where "n" is the number

Microsoft Excel - InterArrivalTimes.xls

File Edit View Insert Format Tools Data Window Help Acrobat

K11   =CHIDIST(K10,7)

	H	I	J	K
1	Bin	Freq, obs	Freq, exp	(F,obs-F,exp)^2 (F,exp)
2	9:00 AM	52	62.5	1.764
3	10:00 AM	66	62.5	0.196
4	11:00 AM	61	62.5	0.036
5	12:00 PM	54	62.5	1.156
6	1:00 PM	68	62.5	0.484
7	2:00 PM	77	62.5	3.364
8	3:00 PM	60	62.5	0.100
9	4:00 PM	62	62.5	0.004
10	Sum	500	500	7.104
11			p-value	0.418

Figure 6.13   Chi-Square Calculation for Customer Arrival Subtotals

of categories. Since there are eight time slots during the day, there are seven degrees of freedom. The p-value of 0.418 is greater than the commonly accepted cutoff of 0.05, indicating that the number of customers arriving during the different times of the day is random. Our conclusion is that even though the histogram of subtotals of customers arriving during each hour of the day showed some variation, it is not big enough to be nonrandom.

If your p-value indicates that your differences are nonrandom, then examine the list of Chi-square values in the K column. Large values indicate a bin where the difference between the observed and expected rates is very large in a relative sense. Examine the large contributions to draw conclusions about the pattern.

The Chi-square test can be applied to data that has been consolidated in any tabular form. When you examine arrival rates where you have more than one factor such as hour of day and day of week, use Minitab to cross tabulate your data and run the Chi-square test. Unfortunately, the test will only tell you if the variation is random or not. It will not quantify the nature of any nonrandom factor.

6.8   ANOVA for Arrival Rates

We are going to analyze the variation in the arrival rate of patients arriving at a walk-in medical clinic. The clinic is open between 7:00 a.m. and 7:00 p.m. during the week and from 7:00 a.m. to 3:00 p.m. on the weekends. The summary of the number of patients arriving every 30 minutes for one week is shown in Fig. 6.14.

The majority of patients arrive early in the day. This is followed by a dip at lunchtime, a brief peak at about 3:00 p.m., before the arrivals trail off after about 5:30 p.m. The rate of arrivals on Saturday is quite high, although the total is less because the clinic closes early on the weekends. The arrival rates on Sunday are about the same as during the week, but the total number of patients is less because of the early closing. We have a few different clusters of behavior:

- There is an hourly variation throughout the day during the week.
- The daily totals change during the week, including the weekends.
- The weekend time profile may be similar for Saturday and Sunday, but different than the weekdays.

The daily variation is probably real because a similar pattern is seen for each day during the week. It is unlikely that the same random variation

Customers Arriving Each 30 minutes

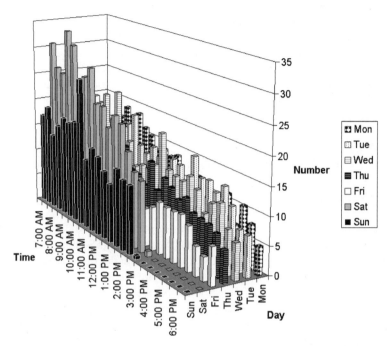

Figure 6.14   Patients Arriving Each 30 Minute Period During the Day Throughout the Week

would be repeated from day-to-day. We could do a Chi-square test on each day to see if the variation is significant, but we will apply a more general technique to this set of data. Strictly speaking, the number of customers arriving during a time interval is a discrete number, but if the numbers of customers arriving within an interval ranges by more than about ten, then the discrete data is pseudocontinuous (Section 5.11.7). Instead of the Chi-square test, we can use the more general *analysis of variance* (ANOVA) to look at the variation during the day. We can perform the ANOVA two ways, first looking at the variation in the patient arrival rate from day-to-day during the week, then again to look at the variation during a particular day.

The Excel function "=WEEKDAY(DATE,2)" was used to determine the day of the week for each date and entered as an extra column in the data sheet. The ANOVA shows that arrival rate on at least one day of the week is different from the others (p-value <0.000). Although the *DayOfWeek* effect is significant, the model explains only 7.5 percent of the variation in arrival rate. This is calculated by the ratio of the *sum of squares* (SS) for the DayOfWeek factor and the SS for the Total (2054/27,117 = 7.6%) This is not a particularly useful result because the size of the effect is small. The clinic does indeed close early on Saturday and Sunday so the daily totals will change, but the rate only changes on Sunday. We could group the day-to-day data in different ways, but it is unlikely that we will gain much more insight into the total variation. We need to look for other factors to explain the remaining variation (Fig. 6.15).

In order to examine the fluctuations during the day, we subgrouped the arrival times into 30 minute bins and retained the flag as an *HourOfDay* factor. The results from the ANOVA examining the variation during the day are shown in Fig. 6.16.

The HourOfDay factor has a significant effect (p-value <0.000), and it explained more variation than the DayOfWeek factor (SS HourOfDay/SS Total = 10,531/27,115 = 39%). Even though this model of average patient arrival rates is significant, it could not be applied to all days of the week: the clinic closes early on Saturday and Sunday. We have also already observed that arrival rates on Sunday are lower than during the rest of the week.

One-way ANOVA: Arrivals versus DayOfWeek

```
Source        DF        SS      MS      F       P
DayOfWeek      6    2054.0   342.3   8.76   0.000
Error        641   25063.5    39.1
Total        647   27117.5

S = 6.253    R-Sq = 7.57%   R-Sq(adj) = 6.71%


                                 Individual 95% CIs For Mean Based on
                                 Pooled StDev
Level   N     Mean   StDev    ---+---------+---------+---------+------
1       72   13.403   4.782                          (-----*----)
2       96   13.260   4.522                          (----*----)
3       96   13.240   5.517                          (----*----)
4       96   11.750   4.389                  (----*----)
5       96   11.365   5.703               (----*----)
6       96   11.687   9.835                  (----*----)
7       96    7.917   6.880   (----*----)
                                 ---+---------+---------+---------+------
                                 7.5      10.0      12.5      15.0


Pooled StDev = 6.253
```

Figure 6.15   ANOVA for Day-Day Variation in Half Hour Arrival Rate

One-way ANOVA: Arrivals versus HourOfDay

```
Source      DF      SS      MS      F      P
HourOfDay   23   10531.6  457.9  17.23  0.000
Error      624   16585.9   26.6
Total      647   27117.5

S = 5.156   R-Sq = 38.84%   R-Sq(adj) = 36.58%
```

```
                                    Individual 95% CIs For Mean Based on
                                    Pooled StDev
Level   N    Mean   StDev    -----+---------+---------+---------+----
 7:00   27  13.593  6.417                      (---*---)
 7:30   27  12.444  5.147                   (---*---)
 8:00   27  14.222  4.577                       (--*---)
 8:30   27  17.111  5.228                            (---*---)
 9:00   27  17.889  4.742                             (---*---)
 9:30   27  15.593  4.466                         (---*---)
10:00   27  18.963  5.626                               (---*---)
10:30   27  18.185  6.349                             (---*---)
11:00   27  14.370  5.062                       (---*---)
11:30   27  12.704  3.969                   (--*---)
12:00   27  12.815  3.617                    (---*---)
12:30   27  11.926  4.075                  (---*---)
13:00   27  10.963  4.155                (---*---)
13:30   27  13.185  5.314                    (---*---)
14:00   27  10.704  3.979               (--*---)
14:30   27  10.741  3.369               (--*---)
15:00   27   9.111  6.891            (---*---)
15:30   27   8.148  6.865          (---*---)
16:00   27   7.926  5.622          (---*---)
16:30   27   7.370  5.719         (---*---)
17:00   27   7.074  5.540        (---*---)
17:30   27   6.926  5.837        (---*---)
18:00   27   5.259  4.935      (---*--)
18:30   27   4.630  3.972    (---*---)
                             -----+---------+---------+---------+----
                              5.0      10.0      15.0      20.0
```

Pooled StDev = 5.156

Figure 6.16   ANOVA for Daily Variation in Half Hour Arrival Rate

It would be best to develop a model that combines the variation during the day with the variation during the week.

6.9 GLM for Arrival Rates

The *general linear model* (GLM) can be considered as a more flexible version of the ANOVA. The technique can accommodate fixed factor levels such as those seen in the ANOVA, and continuous variables such as those seen in multiple regression. We built a model that included the main effects of DayOfWeek and HourOfDay, and the interaction between the two. The output shows that both the main effects and interaction terms are significant.

The ANOVA table shows the same SS for the DayOfWeek factor observed Fig. 6.15, the same SS for the HourOfDay factor observed in Fig. 6.16, plus an additional SS for the interaction between the two main factors (Fig. 6.17). The model explains about 70 percent of the variation in customer arrival rates.

These main effects plots in Fig. 6.18 show the average arrival rate during the day peaks early in the morning, wanes after about 11:00 a.m., and continues to fall off toward the evening. The main effect for the day-to-day variation still shows the "Sunday" effect observed in the ANOVA (Fig. 6.15).

The interaction plots in Fig. 6.19 show the complex interaction between the day and hour effects. In the lower left corner of the plot, the model shows how the customer arrival rate drops to zero on Saturday (day = 6) and Sunday (day = 7) in the late afternoon when the clinic is closed. The upper right corner shows that the days during the week have a similar profile to each other. On the weekends, there is a large peak in arrivals early on Saturday morning, while arrivals on both Saturday and Sunday drop to zero when the clinic closes at 3:00 p.m.

General Linear Model: Arrivals versus DayOfWeek, HourOfDay

```
Factor      Type    Levels  Values
DayOfWeek   fixed       7   1, 2, 3, 4, 5, 6, 7
HourOfDay   fixed      24   7:00, 7:30, 8:00, 8:30, 9:00, 9:30, 10:00, 10:30,
                            11:00, 11:30, 12:00, 12:30, 13:00, 13:30, 14:00,
                            14:30, 15:00, 15:30, 16:00, 16:30, 17:00, 17:30,
                            18:00, 18:30


Analysis of Variance for Arrivals, using Adjusted SS for Tests

Source               DF     Seq SS     Adj SS   Adj MS      F      P
DayOfWeek             6    2053.98    2053.98   342.33  20.72  0.000
HourOfDay            23   10531.62   10087.40   438.58  26.55  0.000
DayOfWeek*HourOfDay  138   6602.96    6602.96    47.85   2.90  0.000
Error                480   7928.92    7928.92    16.52
Total                647  27117.48


S = 4.06431   R-Sq = 70.76%   R-Sq(adj) = 60.59%
```

Figure 6.17   GLM for Half Hour Arrival Rate

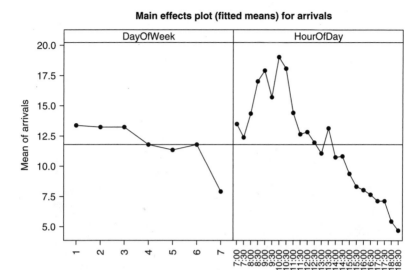

Figure 6.18   Main Effects Plots for HourOfDay and DayOfWeek Factors

Another advantage of using the GLM to examine the data is that the coefficients of the fitted equation can be used to generate predictions given values of the factors. This model fits the observed data well enough that it could be used in a predictive manner to be used in capacity planning. A single week of arrival rates and the predicted rates is shown in Fig. 6.20.

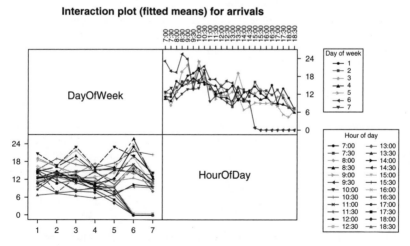

Figure 6.19   Interaction Plot of DayOfWeek and HourOfDay Factors

Patient Arrival Rate

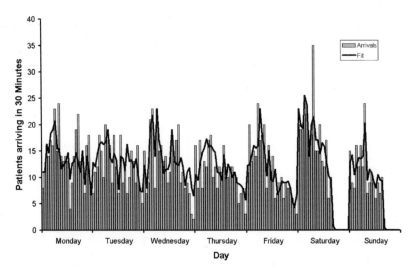

Figure 6.20   Observed and Predicted Customer Arrival Rates for One Week

In this section we have considered that customer arrival rates can change during the days of the week or during the day. The GLM could also be used to examine the variation in hourly arrival rates of patients undergoing different medical procedures.

The same techniques can be applied to examine any other subset of the data. These other factors can come from the list of potential Xs in your data collection plan. Examine your customer arrival rates broken out by different product or service offerings, locations, or customer types.

6.10   Customer Interarrival Times

When we looked at customer arrival rates, we were looking at an average rate for a time interval. If your data on customer arrival times has already been consolidated, then the Chi-square test is the only tool available for you to analyze the customer needs. The data is discrete and even though we could treat it as pseudocontinuous and use ANOVA and GLM, the amount of detail is still a bit limited.

When you have a date/time stamp for each customer arrival, a more useful set of data is customer interarrival times. This is truly continuous data and allows you to conduct a much more detailed analysis. This detail will become very useful when it comes to constructing computer simulations of your process.

The arithmetic difference between successive arrivals gives the interarrival times. The inverse of the interarrival times gives the instantaneous rate of arrival (Equation 6.2).

$$\text{Rate of arrival } = \frac{1}{\text{Interarrival time}} \tag{6.2}$$

Sort your data by arrival time and take the differences between successive customers' arrival times to calculate the interarrival times for each arrival. The first interarrival time is calculated when the second customer arrives and is associated with that customer. For example, the interarrival time for the second customer at our imaginary bank (Fig. 6.11) is 1.949 minutes. This can be expressed as an instantaneous rate of 0.513 customers per minute. Recall that the arrival times were randomly generated with a uniform distribution. We expected roughly the same number of customers to arrive during any time of the day. The Chi-square test verified that the distribution of arrival times was uniform with only random error. What we will gain using the data on interarrival times is a description of the variation in customer arrival rate.

The interarrival times from Fig. 6.11 were copied into a Minitab worksheet to examine the data. The statistical summary is shown in Fig. 6.21. The distribution has a mean of 0.955 arrivals per minute. This might be expected because 500 customers arrived in 480 minutes (0.96 arrivals per minute). Perhaps what is surprising is that the approximately uniform distribution of customers arriving during the day has resulted in a very nonuniform distribution of interarrival times. The distribution ID plots show a good match with the exponential distribution, and the more general version of the exponential, the Weibull distribution (Fig. 6.22).

This is an extremely important result. When customer arrival times are independent, the number of arrivals in equally spaced intervals are uniform and the interarrival times are exponential.

6.10.1 Examples of Interarrival Times

When we were examining different probability distributions in Chap. 5, we found that when the Weibull distribution has a shape parameter of unity, it is identical to the exponential distribution. We use it extensively when examining and summarizing time dependent events.

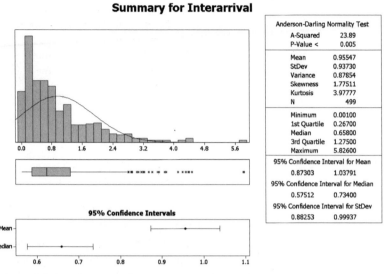

Figure 6.21   Statistical Summary of Interarrival Times

A lot of theoretical work on the arrival times of random events will assume an exponential distribution for convenience. This is a natural consequence of completely independent event arrival times, and works well in some applications. The extra flexibility of the Weibull makes it a better distribution to use when examining customer interarrival times.

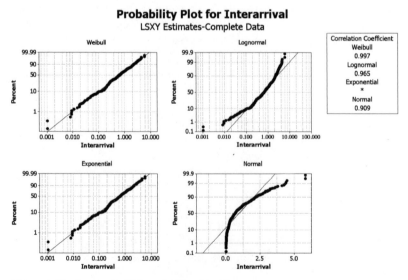

Figure 6.22   Interarrival Times for Random Events

Bank Call Center

We are going to use some data from a call center at a bank. Calls come into a *voice recognition unit* (VRU), where callers may enter their account number, *personal identification number* (PIN), and select a service type using a touch tone phone. The time for incoming calls is recorded to the nearest second. An option is available for potential customers to request information (Fig. 6.23).

The number of inquiries for these new customers is flat between the hours of 8:00 a.m. and 4:00 p.m. and shows no day-to-day variation. The interarrival times were calculated for the ~90 incoming calls each day and expressed as fractions of a 24 hour day. The probability distribution ID plots are shown in Fig. 6.24.

The distribution is best fit using the Weibull distribution with shape parameter of 0.8695 and scale of 0.004183.

Walk-in Medical Clinic

The customer arrival data for the medical clinic was derived from the time stamped on their information forms. The resolution of the data was only to the nearest minute. It was common that more than one form would have the same time stamp, so interarrival times calculated from the raw data would occasionally be zero. It should be obvious that it is not possible for two patients to arrive at exactly the same time, and all interarrival times should be positive. This is a case where the calculation of interarrival times was limited by the resolution of the recording device.

A time recorded as 07:11 could actually be any time from 07:11:00.00 to 07:11:59.99, so we added a random number of seconds ranging from zero to sixty to each arrival time to reflect this indeterminacy. Once the "noisy times" were sorted, we calculated the interarrival times and fit the probability distributions.

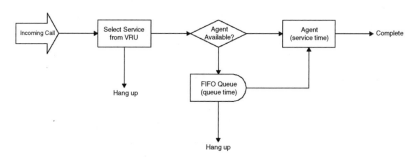

Figure 6.23   Bank Call Center Process Flow for New Customers

Probability Plot for Interarrival time

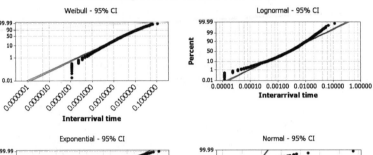

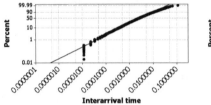

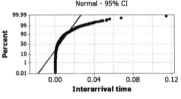

Figure 6.24   Distribution ID Plot for Interarrival Times of Calls From New Customers

Figure 6.25 shows the effect of adding noise to the arrival times before cal-culating interarrival times. The vertical stacks of larger points for the inter-arrival times are caused by a large number of fixed times such as 60 seconds, 120 seconds, and so on. When the random noise was added to the times, the data revealed the distribution shown for the small points. The parameters of

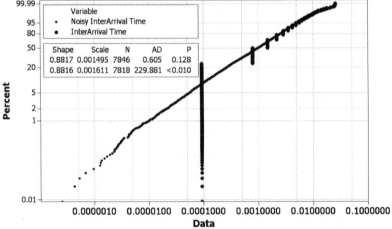

Figure 6.25   Distribution ID Plots for Patient Interarrival Times With Added Noise

the distributions are similar, but the noisy data allows us to detect deviations from the ideal distribution more easily.

6.11   Transforming Weibull Data to Normality

When customer arrival rates repeat from hour-to-hour, day-to-day, or week-to-week in some kind of periodic pattern, it is possible to develop the type of predictive models presented in the GLM section (See Section 6.9). When there is no recognizable underlying pattern to customer arrival times, it is essential to construct a control chart to detect when other, nonperiodic changes occur.

Customer interarrival times are commonly modeled by a Weibull distribution. These data can be mathematically transformed such that they become approximately normally distributed. The transformed data can then be tracked using common control charts. Weibull distributed data is transformed to normality using the equation:

$$y_{\text{transform}} = y^{\lambda} \tag{6.3}$$

where the optimum value for λ is determined by choosing a value that yields a constant variance for all values of y. The standard deviation is given by:

$$\sigma_{y_0} \sim \mu^{\lambda + \alpha - 1} \tag{6.4}$$

When we set $\lambda = 1 - \alpha$, then the variance of the transformed data, $y_{\text{transform}}$, is constant. This is known as the *Box-Cox transformation*. The relationship between the shape parameter of the Weibull distribution and the value of λ is shown in Fig. 6.26.

The interarrival times for new customer calling into the call center were plotted to determine the parameters of the distribution (Fig. 6.27). Rather than using the parameters from the distribution shown in Fig. 6.27 and the relationship in Fig. 6.26, Minitab was used to directly determine the optimum value of for λ the dataset. The output is shown in Fig. 6.28.

The optimum value of 0.23 is very close to a value of zero. Figure 6.26 shows that this is very close to taking the logarithm of the data to achieve normality. This is consistent with the observation that the data are very close to being log-normally distributed (Fig. 6.24). The transformed data are useful for any future analysis where normality is assumed.

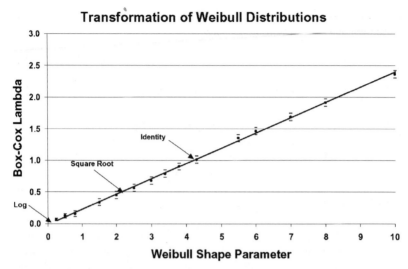

Figure 6.26   The Relationship Between the Weibull Shape Parameter and the Data Transformation Constant

6.12   Control Charts Using Transformed Weibull Data

The arrival rate data for the walk-in clinic was previously analyzed using ANOVA where it was found that the DayOfWeek effect accounted for only a small, but significant fraction of the variation in the data. It would be interesting to see if the "Sunday" effect could be detected using control charts

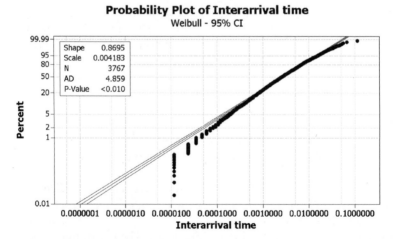

Figure 6.27   Weibull Distribution Parameters of the Call Center Interarrival Times

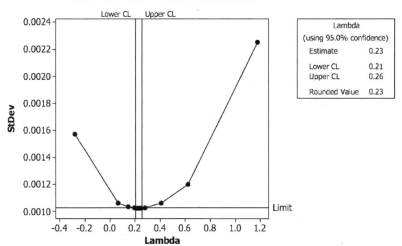

Figure 6.28 Determination of the Optimum Value of Lambda for New Customer Interarrival Times

of the interarrival times. The data for one week of arrivals was transformed using the Box-Cox transformation (Fig. 6.29).

Figure 6.30 shows the Xbar-S chart of the transformed interarrival time data for seven successive days during one week. The sample mean chart (top) shows that the average interarrival time dropped on day six (Saturday). This is consistent with the observation that Saturday is a busy day. This chart also shows that on day seven (Sunday), the average interarrival time is back in control. Any out-of-control condition detected in the sample standard deviation chart (bottom) would indicate a day which is much more consistent or inconsistent for patient arrivals.

The GLM analysis produced a model of expected customer behavior based on one month of data where the multiple observations of arrival times for each day of the week were combined. The data in this control chart is for only one week. The difference in the conclusions about the weekend effect show it is still possible to experience variation not explained in the model.

It is possible to construct a control chart where the raw data is the difference between the actual and predicted customer arrival rates. This chart would indicate whether changes to the model are required. Control charts can quickly detect both short-term and long-term changes in customer behavior in an objective manner without having to constantly perform the GLM analysis.

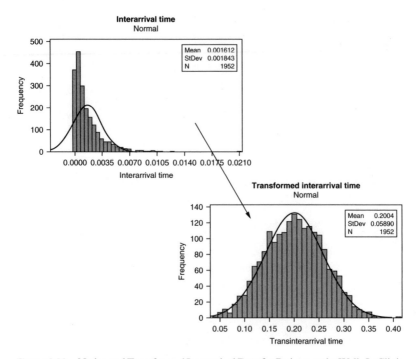

Figure 6.29 Native and Transformed Interarrival Data for Patients at the Walk-In Clinic

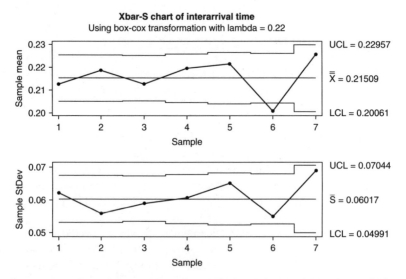

Figure 6.30 Control Chart of Transformed Patient Interarrival Data for the Walk-In Clinic

6.13   Nonparametric Tests

The two-sample *t*-test, ANOVA and GLM work well if the distribution of errors is approximately normal, or at least symmetric. More sophisticated statistical tests have been developed to deal with other distributions, outliers, or small data sets. These nonparametric, or "distribution free" tests are quite robust with respect to dealing with nonnormal data, but are not as sensitive as the parametric tests; they might not detect differences that are present.

Nonparametric tests are usually the best ones to use if you are using untransformed data. The two-sample Wilcoxon rank sum test or Mann-Whitney test can be used when testing the equivalence of the medians of two subgroups when each subgroup has approximately the same nonnormal distribution. The Mood's median test and the Kruskel-Wallis test are the nonparametric equivalents to the ANOVA. The Mood's median test is best when the subgroups have different, nonnormal distributions, and is robust against outliers, but is not very sensitive. The Kruskel-Wallis is better when the subgroups have approximately the same distribution, is sensitive to small changes in medians, but outliers can cause problems.

The Minitab help file summarizes many of the characteristics of the non-parametric tests and the circumstances where they may be the best choice. In general, we do not recommend using the nonparametric tests. There is a wealth of information that can be gained by understanding why your data is nonnormal. Cycle times, execution times, and customer arrival times show distributions which are nonnormal, but well- understood. Understand and characterize the distributions first, then transform them it to perform statistical tests.

The Mood's median test is useful to verify whether the medians are different for different subgroups because the summary of *P5-P95* span includes the median. The median is also the single parameter that never changes when data are mathematically transformed.

6.14   Analyzing Production Capacity–Execution Time

We have completed the first step in aligning the customer needs to the business requirements. We now have to spend some time examining how your business responds to these needs by examining the time it takes to provide service to these customers. Execution times for manufacturing operations can be quite reproducible. The time taken for a repetitive machining operation will generally be a certain minimum time and have some variation about the mean. Transactional systems and service processes tend not to be so reproducible.

We will be examining the execution time for assisting new customers contacting the call center of a bank, first seen in Section 6.10.1 and Fig. 6.23. Recall that the customers first enter data into a VRU and select a service type using a touch tone phone. Once identified as new customers, they are entered into a *first in, first out* (FIFO) queue. The calls are routed to an agent when one becomes available. The beginning and end times for services are logged. The execution time for service is the elapsed time starting when the agent first talks with the customer and ends when the call is complete. We are not considering any routing or waiting time at the moment.

The statistical summary of the data (Fig. 6.31) shows that execution time is highly skewed. Using only the average and calculating business capacity based on that single number would mask the observation that there is a great deal of variation in call length.

The distribution is clearly nonnormal; the mean call is 121 seconds, but the longest call is over one hour long. The distribution ID plot shows an excellent fit with a log-normal distribution (Fig. 6.32).

Summary for Service Time

Anderson-Darling Normality Test	
A-Squared	445.37
P-Value <	40.005
Mean	121.06
StDev	177.36
Variance	31457.90
Skewness	7.842
Kurtosis	117.681
N	3798
Minimum	1.00
1st Quartile	41.00
Median	71.00
3rd Quartile	137.00
Maximum	4258.00

95% Confidence Interval for Mean	
115.42	126.70

95% Confidence Interval for Median	
69.00	74.00

95% Confidence Interval for StDev	
173.46	181.45

Figure 6.31   Statistical Summary of Call Center Service Time

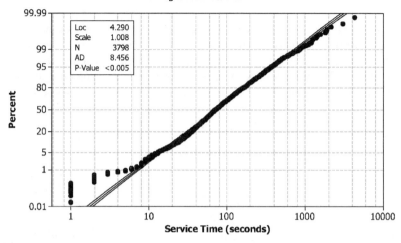

Figure 6.32   Log-Normal Probability Plot for Call Center Service Time

The deviations from the ideal, log-normal distribution are caused by a small proportion of calls from 1 to 5 seconds in length. These calls are mostly from people who have reached the new customer agents by mistake. The data was transformed to normality using the Box-Cox transformation with lambda equal to 0.04 and stored for later use.

6.15   Testing for Subgroups of Execution Times

When service is being provided, it is common that there will be different subgroups of specialized workers with differing capacity for different types of service. When we were examining customer arrival rates for the walk-in medical clinic and the bank call center, we examined the differences between different subgroups using a variety of statistical methods. We must do the same with execution times to discover sources of variation in service times. The two measures are central tendency and variation. These can be characterized using mean and standard deviation for approximately normally distributed data, and median and *P5-P95* span for probability distributions for nonnormal data.

We segmented the service time data for the call center by experience of the agent. This factor came from discussions with the process owners who proposed leaving out one particular agent's calls from the analysis, feeling that his level of experience would bias the data. The probability plot of execution time was repeated for the two subgroups (Fig. 6.33).

The distribution of service times for the trainee is to the right of the other agents in the figure. Since the incoming calls are routed randomly to whoever is available, the difference in service time for the trainee is likely to be caused by the difference between agents and not some other factor. The distribution analysis has characterized the large variation within each group in a very consistent manner. The difference between the two distributions show there is a systematic difference in the way calls are serviced by the two subgroups of agents.

We have some choices in the statistical tests to verify if the differences in central tendency and variation are significant. We could use tests that do not require normally distributed data (nonparametric tests), or transform the data to normality and then conduct parametric tests. When data is nonnormal in a well-understood manner, we usually transform it first before running the tests. If you have any doubt about your own data, run the tests using both the untransformed and transformed data. We have found the conclusions are rarely different.

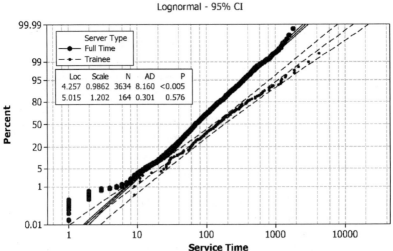

Figure 6.33   Log-Normal Probability Plot for Call Center Service Time Broken Out by Experience Level

The test for differences in variation between subgroups of data is termed a *test for equal variances* in Minitab. Of the two statistical tests performed simultaneously, the Levene's test is a more conservative test and especially superior when the input data is nonnormal. It should be examined whether the data has been transformed or not.

The p-value from the output (<0.000) indicates that the difference in variation between the two subgroups is significant (Fig. 6.34). The range of call length for the trainee is wider than it is for more experienced workers. Note that the box-plots for the subgroups in Fig. 6.34 are approximately symmetric, indicating the success in transforming the highly skewed data shown in Fig. 6.31.

Since we have only two members in the subgroup, we can use the two-sample *t*-test to check for difference in central tendency for the transformed data. There are two slightly different ways of conducting the two-sample *t*-test depending on whether the variances of the subgroups are equal or not. We have just found from the Levene's test that we should be more conservative in running the test by not assuming equal variance of the subgroups. If we had more than two members in the subgroup, we would run the ANOVA test on the transformed data. The box-plot of the two-sample *t*-test on the transformed data is shown in Fig. 6.35.

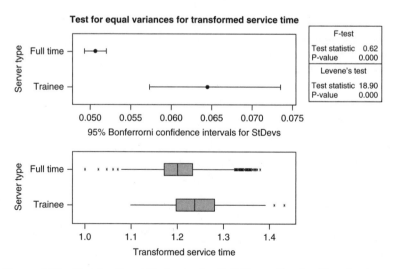

Figure 6.34   Test for Equal Variances for Call Center Service Time by Agent Experience

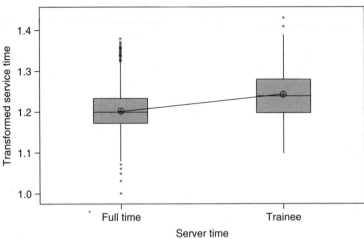

Figure 6.35   Box-Plot for Mean Transformed Call Center Service Time by Agent Experience

The mean for the trainee is higher than the mean for the experienced staff members. The diagnostic output from Minitab is shown in Fig. 6.36.

The p-value (<0.000) indicates that the difference in mean transformed service time is significant. Although we have proved the significance of the differences in service time for the trainee versus the experienced staff, we still need to summarize those differences in a manner that is easily understood without all the statistical detail.

```
Two-sample T for Transformed Service Time

Server Type    N    Mean    StDev   SE Mean
Full Time    3634  1.2017  0.0506   0.00084
Trainee       164  1.2421  0.0644   0.0050


Difference = mu (Full Time) - mu (Trainee)
Estimate for difference:  -0.040313
95% CI for difference:  (-0.050380, -0.030246)
T-Test of difference = 0 (vs not =): T-Value = -7.90  P-Value = 0.000
DF = 172
```

Figure 6.36   Two-Sample T-Test for Mean Transformed Call Center Service Time by Agent Experience

6.16   Summarizing Subgroups of Execution Time

The preceding analysis has confirmed that the server type is an important factor in the mean and variance of service time. Similar analyses may be performed for each one of a number of proposed factors. These candidate factors will come from your data collection plan and were listed when the team brainstormed all the potential factors that may affect service time. A summary of the significant results should be made to show the differences in median and span for the subgroups.

The Excel "volume-high-low-close" chart type can be used to summarize these results. The chart for the service type broken out by the "server type" factor is shown in Fig. 6.37.

The median call length for the trainee was 145 seconds, compared to 71 seconds for the full time staff. The difference in *P5* shows that the trainee takes about twice as long as the rest of the staff even for the short calls.

6.17   Measuring Customer Patience

The call center for our bank handles the incoming calls for all kinds of services. We are going to measure the patience of only the new customers who enter the VRU and choose the option to talk with someone to get

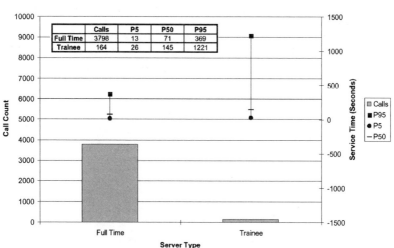

Figure 6.37   Median and *P5-P95* Span of Call Center Service Time by Server Type

information about services. If an agent is not immediately available, the customers enter a queue and wait until there is one. There is a periodic message played asking the customer to wait rather than hang up and call again. There are two possible outcomes when they wait, they can remain on the line until they are served, or they can hang up after they lose their patience. In the sample under study, over half the customers hung up from the queue before talking with an agent. It is of great interest to quantify the time dependence of this behavior (Fig. 6.38).

In the field of reliability engineering, this is called *failure analysis*. One way to measure customer patience would be to have all customers remain in the queue and measure the time until they each hang up. This is not practical, polite, or good business practice. Neither can we simply select all customers who hang up and analyze that subgroup. The problem with the latter approach is that there may be customers that would normally hang up after 10 seconds, but reached a server beforehand and would not be counted. The data must be segmented to recognize that some customers stayed on the line long enough to hang up, while others were removed from the queue by being served. What we know about the customers who received service is that they did not hang up in at least X seconds when they got service; they may have hung up a few seconds later. This is termed *right-censoring* and we have to flag data to indicate this fact. Censoring was not necessary when we were analyzing service time because all customers received service.

We extracted the data for queue start times and end times for new customers over one month and calculated queue time. The systems logs whether the customer reached an agent (flagged "AGENT") or hung up before talking with someone (flagged "HANGUP"). We will determine the probability distribution of the customer patience using the *parametric distribution analysis with right censoring* procedure in Minitab. Calls flagged as "AGENT" will

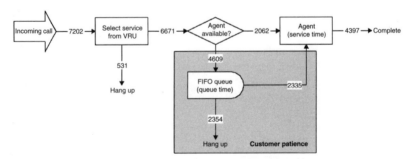

Figure 6.38   Measuring Patience of New Customers at the Call Center

be censored. Minitab recognizes that these customers were pulled out of the queue by being answered (right-censored at that time) rather than stayed in the queue to eventually hang up (failure at that time). The distribution fits the three-parameter Weibull distribution quite well (Fig. 6.39).

The 0.92 second threshold parameter means there is about one second delay before the time dependence of patience starts. The probability plot in Fig. 6.39 can be used to determine the percentage of customer who will hang up after a given time. For example, after 100 seconds of hold time, about 40 percent of customers will have hung up.

A *survival plot* is another way of showing the time dependence of customer patience. The plot shows the percentage of customers who will remain on hold, given that they have waited a certain period of time (Fig. 6.40).

A reasonable objective might be to determine what percentage of customers would be willing to wait, say 400 seconds. Figure 6.40 shows that only about 20 percent of new customers will wait that long. A *hazard plot* indicates the rate at which customers hang up given that they have stayed on hold for a given time. This plot indicates that the rate at which customers hang up is large to begin with, then levels off after about 100 seconds (Fig. 6.41).

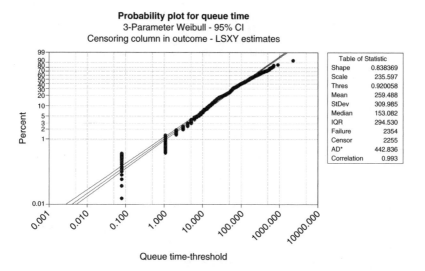

Figure 6.39 Probability Distribution of Call Center Customer Patience with Right Censoring

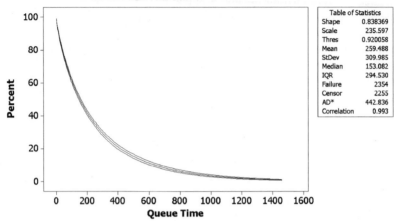

Figure 6.40   Survival Plot Showing the Relationship Between Hold Time and Patience

This same analysis can be applied to other subsets of customer subgroups for the call center. Priority customers, stock exchange information, internet banking, financial planning services, mortgage, and loan processing are only some of the types of subgroups that could be analyzed in the same manner.

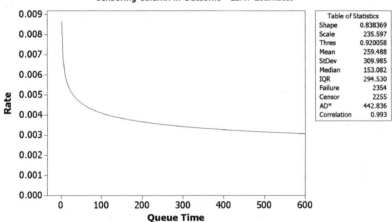

Figure 6.41   Hazard Plot of Call Center Customer Patience

Keep in mind that this analysis is telling you about the customers who are not satisfied with the length of time they must wait to contact your bank's service. If this was your lean Six Sigma project, you must return to this matter during the Improve section. If you replaced the "on-hold" message with an effective message redirecting certain types of calls to the bank's internet site, then you would expect a higher number of people hanging up after the message was played. This new "on-hold" message could even be played for half the incoming calls during a test period to facilitate the comparison.

6.18 Delay Time

The preceding examples using data from a call center showed the wealth of information that can be obtained from a data set rich in information. It is possible to do extensive analysis of customer arrival rates, queue delay time, abandonment behavior, and service cycle times. These elements can all be combined to determine how they relate to one another.

Look at the process of managing incoming calls shown in Fig. 6.38 from a "lean" perspective. About 30 percent of the calls go directly to an agent, but the remainder have to wait. The calls sitting in the "on-hold" queue are *work in process* (WIP). The danger here is not so much that the cycle time is increased because of the backlog, but more that calls placed in the WIP queue result in about a 50 percent "scrap" cost from customers hanging up from the queue. This same behavior is seen in other transactional systems. Many mortgage and insurance applications never end up getting signed because the delays are so long or unpredictable that customers have purchased the service from your competitors.

It is extremely rare that a transactional system operates anywhere near a flow system. Each process step acts like a combination of a queue and a process step in a complex relationship. A corporate focus on maximum utilization of resources and cost reduction is a very difficult mind-set to break and is the principal cause of customer dissatisfaction shown by abandonment.

Queuing theory is the discipline that explores the relationship between customer arrivals, providing services, and waiting for service. There are two approaches to studying systems:

1. Analytic–Make assumptions about the arrival rates and execution times and use standard equations to determine parameters of interest, such as average time in the queue or average length of the queue. These tend to be restricted to simple, ideal systems with limited flexibility.

2. Simulation–Determine the types and parameters of distributions using historical data and use Monte Carlo simulations to determine the behavior of the system and the effect of changing input parameters or configurations. This approach can handle all kinds of situations that are not possible using the analytic approach. Customers might not enter the queue if they think it is too long (*balking*), they may switch from one queue to another (*jockeying*), or leave the queue after waiting too long (*reneging*). Customer priorities might also change once they are in the queue.

We favor the simulation approach and will use it to explore the influence of different factors in these systems.

We used the simulation software, ProcessModel (See Appendix B), to construct a simple system. A single person processes application forms as they arrive in an input queue. The worker eventually completes all documents. The software allows for gathering data from the process as the simulation is running. We can use the model to investigate some elementary properties of queuing systems and experiment with a few different configurations (Fig. 6.42).

The arrival times of the application forms was assumed to be random. We have already investigated a number of customer arrival processes and found them to have a Weibull interarrival time close to the ideal exponential. We specified this to the model. In manufacturing processes, it is common that the execution time of the process is quite consistent. This is usually not true in transactional systems, we have already seen Weibull and log-normal service times in this chapter. For the purposes of this investigation, we used an exponential distribution.

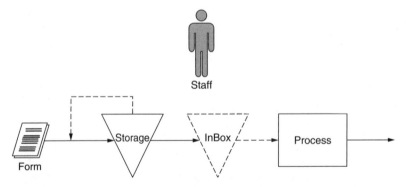

Figure 6.42   Simple Process Simulation with One Worker and One Queue

In a perfect flow system, each piece will flow into the next step of the process at the pace of the entire system—the "takt" time. When this is true, no inventory builds up. One problem we see immediately in this ideal system is that requests do not come at regular intervals. You can easily observe that at a bank or medical clinic, customers will arrive independently. Even if patients arrive at about 50 people per hour, there may be a 1 minute period when no one arrives followed by another 1 minute period when three people arrive. Even if patients are served at exactly 1 per minute, there will still be some variation in total cycle time caused by the variation in waiting time. There will also be some time periods when all patients have received service, the queue is empty, and the workers are idle.

The model system was run a number of times with different loading of the worker. The relationship is not linear, as the loading of the worker goes beyond about 90 percent, the average number of arrivals waiting in the queue increases dramatically, with a consequential dramatic increase in cycle time (Fig. 6.43). This figure shows only the average of execution time and cycle time under different loading conditions. When the queue is heavily loaded, there is a great deal of variation in the number of people waiting (Fig. 6.44).

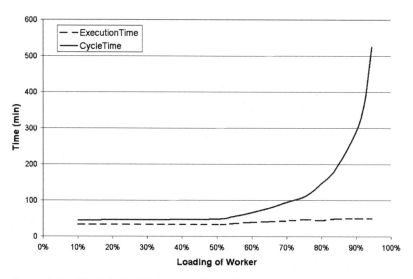

Figure 6.43   The Relationship Between Worker Load and Customer Contact Time

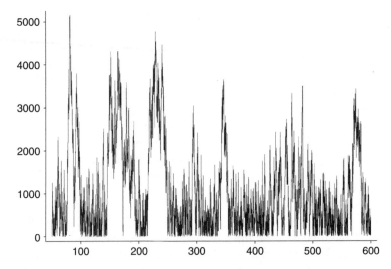

Figure 6.44   History of Cycle Time for a Random Regime at 90 Percent Loading

We ran the model at 90 percent loading to emphasize the effect of different ways of handling the priorities in the queue.

Think of someone placing application forms in an "In" box on your desk for you to process. You could handle your backlog by always taking the oldest application form at the bottom of the pile. This is the way call centers handle calls, termed FIFO. Another way would be to always take the newest application from the top of the pile. This is termed *last in, first out* (LIFO). You could also just pull one out at random and complete the transaction.

The history of the cycle time for the random regime shows a great range in values (Fig. 6.44). The queue length can grow in length and remain that way for some time. The majority of the time, there are about 20 items in the queue, but occasionally the queue can contain many hundreds of items. The history plots of the FIFO and LIFO regimes look similar. The data does not appear normally distributed.

The Weibull distribution plot for the cycle time data for the random, FIFO, and LIFO regimes is shown in Fig. 6.45. The FIFO and random processes have a similar profile, whereas the LIFO process has a "kink" in the middle of the graph and has cycle time values that extend to extremely large values. This is because the input queue rarely gets small enough to get the oldest

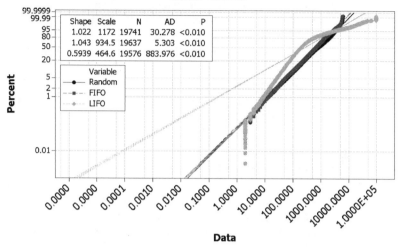

Figure 6.45   Weibull Distribution Plot for Cycle Time at 90% Load with Different Queue Regimes

applications at the bottom of the pile. The summary for the three sets of data is shown in Fig. 6.46.

The numbers of applications processed and average cycle time is essentially the same for the FIFO, LIFO, and random regimes. The *P5* and median (*P50*) for the LIFO queue is quite a bit shorter than the other two regimes. This is a consequence of the "kink" in the probability plot. The median has been moved to a lower value for the LIFO regime, but at the cost of having some extremely large values for cycle time for the longest applications and larger span for the entire data set.

		Single Queue Handling		
		Random	FIFO	LIFO
Cycle Time	P5	61	54	27
	P50	802	664	190
	P95	3327	2770.2	3294
	Average	1087.0	1055.7	1075.2
Inventory	Total	19833	19827	19803
	Average	10.9	10.6	10.8

Figure 6.46   Performance of Single Queue with Different Handling Regimes

There are a few general conclusions:

- Different ways of handling the input queue can affect the measures of central tendency (*P50*-median) and variation (*P5-P95*).
- What may appear to be an improvement in one metric may result in the deterioration of another. We must define our measure of efficiency carefully.
- Extremely long queues can be generated when a system is pushed to near 100 percent usage. An increase of only 5 percent can double the cycle time for heavily loaded systems.

6.19   The Consequences of Changing Priorities

The model of the simple queue can be used to investigate the consequences of personal performance metrics. We have constructed a model where the worker can have some choice in which applications are selected for work.

The data summary shows that there is an average of about 11 application forms in the input tray at any one time. We have already found that execution times can have a wide range in transactional systems. An example might be a person who is researching pricing information for multiple orders. The processing time for each order can vary widely depending on the number of line items on the order. Most experienced workers can estimate the time it will take to process a particular order. With this in mind, we investigated what occurs when workers try to maximize their own throughput of orders by choosing to process shorter orders before longer ones.

We altered the process to include the sorting of incoming applications into 10 roughly equal groups of an "In" box. The 10 groups vary from extremely large orders to very small ones. Whenever the worker has finished with an application, the worker selects the next application from one of the 10 slots of the "In" box, and works on it until it is complete (Fig. 6.47).

The three scenarios were:

- Shortest First–of the applications in the "In" boxes, choose the one with the shortest estimated execution time.
- Longest First–of the applications in the "In" boxes, choose the one with the longest estimated execution time.
- None–choose one of the applications in the "In" box at random.

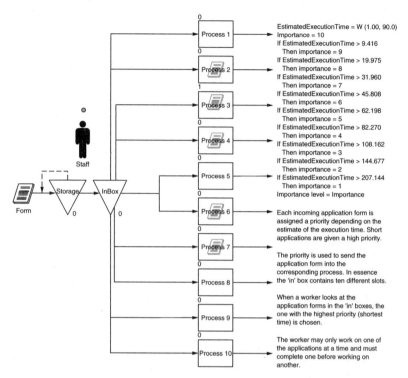

Figure 6.47   Application Form Processing with High Priority Given to Short Applications

The history plots for cycle time look similar to those seen in the simpler case (Fig. 6.44). The probability plots are shown in Fig. 6.48. The *shortest first* plot is similar to the one seen for LIFO in Fig. 6.45. The effect on metrics is similar, the P5 and P50 (median) are substantially shorter than the *none* regime, but the gains in the short and medium range are offset at the expense of the very long. The regime of *longest first* has little effect in the short and mid range, but the effect on long cycle times is enormous. Both the P95 and average are considerably longer. The inventory levels are the lowest for *shortest first* and highest for *longest first* regimes (Fig. 6.49).

The determination of the optimum queuing regime is extremely complex. It depends on the metric of interest, whether it be economic value, cycle time, or minimum span.

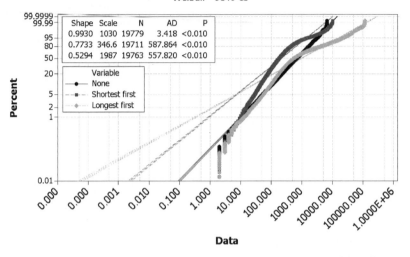

Figure 6.48   Weibull Distribution Plot for Cycle Time at 90% Load with Different Handling of the Queue

6.20   Leveling Arrival Times and Execution Times

When the single process queue was loaded at about 90 percent, the effect of changing the priorities of the items in the queue resulted in changes to the span of the process. The model was run without changing the distributions of the interarrival times or the execution times. *Load leveling* is the term given to various techniques of altering the distribution of arrival times, while *standard setup* alters the distribution of execution times. To investigate the relationship between these two, we ran ProcessModel on the single queue process with four combinations of interarrival times and execution times. The distributions were changed to be either exponential or normal using the

		Priority Regime		
		None	Shortest first	Longest first
Cycle Time	P5	110	93	105
	P50	710	240	544
	P95	2465	1759	11358
	Average	950.3	492.1	2605.5
Inventory	Total	19736	19758	19722
	Average	9.5	4.9	26

Figure 6.49   Performance of Changing Priorities According to Execution Time

		Shape	Scale	Mean
Arrival	A1	1.0	100.0	100.0
	A2	3.5	111.7	100.0
Execution	E1	1.0	75.0	75.0
	E2	3.5	83.3	75.0

Figure 6.50   Parameters of Interarrival Time and Execution Time to Determine the Effect of Changes in Distributions

Weibull shape parameter. The parameters were constructed such that the resource utilization was 75 percent in all cases (Fig. 6.50). The simulations were run, data gathered, and summarized in Fig. 6.51.

The same number of items have been processed in each of the four scenarios, but the performance is quite different using other metrics. The span shows that the worst scenario is A1E1, where both the interarrival time and execution times are exponential. If we had an improvement effort directed toward making the execution time more consistent (A1E2), the effect would be to narrow the span on cycle time by decreasing the *P95* and median at the cost of increasing the *P5*. If the customer arrival times were made more consistent by load leveling, it would have about the same effect as changing the execution times (A2E1). By far, the largest change would be to make both customer arrivals time and execution times more consistent (A2E2) (Fig. 6.52). This also results in the lowest inventory levels of the four scenarios (Fig. 6.51).

There is an important characteristic when the execution time is normally distributed—the probability plot of cycle time has a distinct "kink." The location of the "kink" depends on the relative amount of items that pass

		A1E1	A2E1	A1E2	A2E2
Cycle time	P5	18	11	54	40
	P50	217	126	157	86
	P95	972	509	527	159
	Span	954	498	473	119
Inventory	P5	0	0	0	0
	P50	2	1	1	0
	P95	11	5	6	1
	Span	11	5	6	1
Count	N	19640	19629	19629	19629
No Waiting		24%	42%	24%	61%

Figure 6.51   Summary Statistics for the Scenarios with Different Distributions of Interarrival and Execution Times

Probability Plot of A1E1(cycle), A2E1(cycle), A1E2(cycle), A2E2(cycle)
Weibull - 95% CI

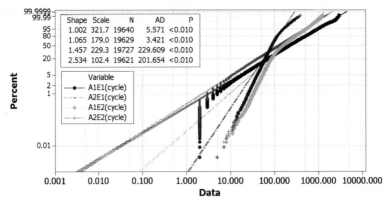

Figure 6.52 Cycle Time Probability Plots with Different Arrival and Execution Scenarios

through the process without delay. The kink for the A2E2 process is at a higher percentile than the A1E2 process. Compare with the percentages of items that are processed with no waiting shown in Fig. 6.51. The location of the "kink" demarcates the items that have passed directly through the process from items that waited in some kind of delay state.

6.21 Calculation of Transactional Process Efficiency

The call center gave us some crude data for the estimation of the efficiency of the process for handling incoming calls. Note that our definition of efficiency is strictly in terms of the value stream from the customer viewpoint. We have not referred to any of the more traditional, internal uses of the word, *efficiency*. We have not discussed such things as resource usage, idle time, or time on call for operators, and such. We are trying to measure the proportion of transactions that proceed as part of a flow systems versus those that are delayed either by being placed in lower priority or wait in a queue because the system is backlogged.

We could calculate efficiency using the number of handled calls versus the total calls, or the number of quick calls per day versus total calls per day, but the emphasis in lean Six Sigma is the flow of individual transactions. The transaction for customer in the call center is the total cycle time of the call. Some calls are directed to an agent immediately, while others must wait because they are in the waiting queue—WIP. Although this is a natural consequence of a limited number of servers with a varying number of incoming calls, we still must quantify it somehow.

An example of the calculation is shown for the sourcing step of the quote to shipment process for our large equipment manufacturing facility. The process step entails that the sourcing staff call the subcontractor and verify the price and delivery time for a component of an order. The cycle time measured is the time to obtain the information about the part, not the part itself. This information is passed on and assembled into a quote to be presented to the customer. The statistical summary is shown in Fig. 6.53. The probability plot for cycle time for this step is shown in Fig. 6.54.

The "kink" in the cycle time plot, first seen in Fig. 6.52, shows that the process has two distinct time dependent components. The dotted lines in Fig. 6.54 extrapolate the two components of the process to an intersection to identify which items are classified as belonging to the fast process (left side) and which are classified as belonging to the slow process (right side). The two dotted lines cross at about the 75th percentile: about 70 to 80 percent of the line items proceed very quickly, while the remaining ~20 to 30 percent proceed slowly and extend out to very large cycle times.

We calculated a hazard plot using the same data and saw that it showed a similar shift from a high to low rate of execution (Fig. 6.55). The rate is very

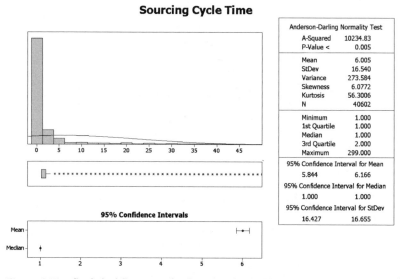

Figure 6.53   Statistical Summary for Sourcing Cycle Time

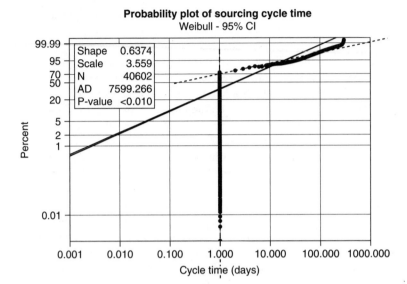

Figure 6.54   Probability Plot for Sourcing Cycle Time

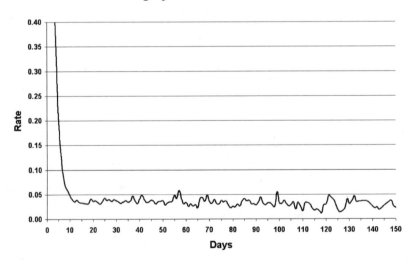

Figure 6.55   Hazard Plot of Sourcing Cycle Time

Descriptive Statistics: Sourcing Cycle Time

Variable	Fast/Slow	Total Count	Mean	StDev	Minimum	Maximum
Sourcing Cycle T	Fast	32317	1.0929	0.2902	1.0000	2.0000
	Slow	8285	25.166	29.652	3.000	299.000
	All	40602	6.005	16.540	1.000	299.000

Figure 6.56 Descriptive Statistics of Fast and Slow Processes for Sourcing Cycle Time

high for short cycle times, but drops to a very low rate after the line items are older than a few days. If we use a cutoff of 3 days as the division between fast and slow, then a tabulation of the cycle times give a proportion of 80/20: fast/slow (Fig. 6.56). A further conclusion from the hazard plot (Fig. 6.55) is that after 3 days, the items are selected at random to be executed.

The measure of process efficiency is the amount of time spent in the fast process divided by the total time spent in the process.

$$\text{Efficiency} = \frac{(p_{\text{fast}}) \times (\mu_{\text{cycle,fast}})}{\mu_{\text{cycle,total}}} \tag{6.5}$$

Substituting values from Fig. 6.56,

$$\frac{\left(\frac{32317}{40602}\right) \times 1.0929}{6.005} = 14.5\% \tag{6.6}$$

This value is a bit surprising, given that 80 percent of the items proceed through the process relatively quickly. The problem is that the 20 percent of items that take longer than 3 days, take very much longer than 3 days. It is for this reason that we do not state that the process is 80 percent efficient. This metric shows the two ways of increasing the efficiency of the process:

- Decrease the proportion of items in the slow process.
- Decrease the average time of items in the slow process.

6.22 Analysis of Transactional Process Efficiency

The hazard plot of the cycle time data from the sourcing department showed the rate of execution for the process (Fig. 6.55). When line items became older than about 5 days, the rate for execution did not change. The items in the slow queue were executed at random—a 15 day old item had the same probability of being executed as a 120 day old line item. This has important considerations when it comes to screening the factors that may influence whether an item proceeds via the fast or slow process. Since the rate for the slow process was not affected by the elapsed cycle time beyond about 5 days, regression analysis is of little use in testing the significance of a potential factor.

In order to determine the drivers of process efficiency, we must do something counterintuitive with the cycle time data. We must convert it from continuous to discrete data. It is fairly straightforward to divide the cycle time data into two categories based on the probability plot.

1. Plot the data using a Weibull distribution.
2. Identify the approximately straight sections of the plot on either side of the "kink."
3. Draw a line on each section of the plot and note the coordinates of the intersection.
4. Use the calculator function in Minitab to flag all data points to the left of this intersection as "slow" and points to the right as "fast."
5. Use the *descriptive statistics* function in Minitab to identify the mean value and number of points in each class.
6. Calculate the process efficiency using Equation 6.5.

6.23 Binary Logistic Regression

When we flagged the cycle time data as "fast" or "slow" in Section 6.22, we defined the response variable as discrete. The statistical tools for analysis are the Chi-square test and binary logistic regression. If our potential factor was discrete, such as sales region, we could cross-tabulate the fast/slow factor against the sale region factor and run the Chi-square test against the result. As we found when using the Chi-square test for customer arrival rates in Section 6.2, the Chi-square test was rather crude and not very useful for quantifying the effect of a factor. Our preferred tool was the GLM in those circumstances. We find it to be applicable in a wider range of data types with the flexibility to build sophisticated models for quantifying the results. Binary logistic regression is the preferred tool for discrete responses for similar reasons.

Binary logistic regression is a specialized version of the GLM where the response variable is transformed to the logit or log-odds ratio. Given the proportion of the population, the logit function is

$$\text{logit}(\pi) = \log\left(\frac{\pi}{(1-\pi)}\right) \qquad (6.7)$$

The tool will give output in the form of odds ratios for different values of input factors.

6.23.1 Insurance Underwriting

When a customer wishes to purchase a life insurance policy, he will contact an agent to prepare and sign an application. The application is sent to a central facility for underwriting—the process of gathering the relevant information, rating the policy, and preparing the payment schedule. The policy is then printed for presentation to the customer. This process may be repeated if the coverage is revised or information is incorrect. One of the longest steps in the process is the delay in receiving information before an underwriting decision is made. This information could come from past insurers, medical personnel, or other professionals. The summary for a sample of cycle time data from the *information request* step of the process is shown in Fig. 6.57.

The cycle time for information request shows the familiar peak at short values with a long tail out to extremely high cycle times. The probability plot indicates that about 35 percent of the applications take about one day, while the remainder stretch out to over 100 days. The policies were flagged as "fast" if they took one day, and "slow" otherwise (Fig. 6.58).

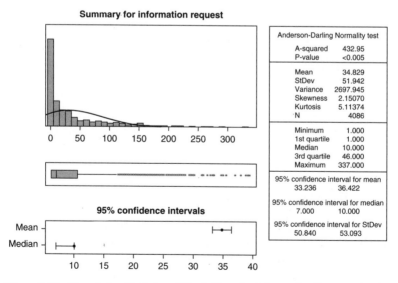

Figure 6.57   Descriptive Statistics of Cycle Time for Information Requests During Underwriting

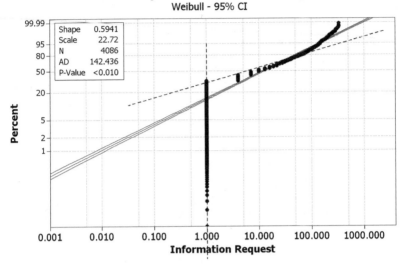

Figure 6.58 Probability Plot of Cycle Time for Information Requests During Underwriting

Rather than merely characterize the process efficiency, we wished to compare the behavior of the different sales regions. The probability plot was repeated, broken out into the three sales regions (Fig. 6.59).

This showed that the central sales region not only had a greater number of policies, but also had the highest proportion of the slow process. The Chi-square analysis showed a significant difference in the proportion of the fast process in the different regions.

A binary logistic regression model was constructed that included the total coverage amount of the policy. We suspected that the agents were processing information requests for the larger policies before the smaller ones. We added an interaction term between policy amount and sales region to investigate whether the policy size had a different effect in different sales regions. The output is shown in Fig. 6.60.

The odds-ratios refer to the response, "slow" as the "event." We have a total of 2690 "slow" events out of a total of 4078. The logistic regression table has entries for coefficients, log-odds ratios, Z and p-values. You should notice there is no entry for the "central" region. This is the reference value of the "region" factor. The coefficients for "east" and "west" are compared

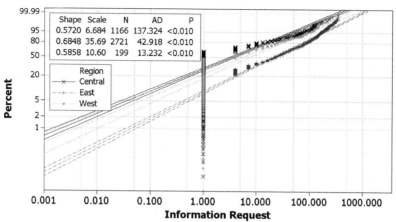

Figure 6.59   Probability Plot of Cycle Time for Information Requests Broken Out by Sales Region

to the reference odds for "central." The coefficient of 1.74829 is the log of p(slow)/ p(fast) for the eastern region. This value is also represented as an odds ratio to the right. The interpretation of this line is, "The odds of an item being processed slowly (reference event) increase by 5.74 times, if the item comes from the East versus coming from the Central sales region." The interpretation of the following line is, "The odds of an item being processed slowly increase by 1.87, if the item comes from the West versus coming from the Central sales region."

The coefficient for "amount" is zero, and the corresponding odds ratio is unity. The coefficient for the interaction terms between "amount" and "region" are also zero. The p-values indicate that the "region" effect is the only significant one. The goodness of fit p-values are all greater than 0.05, indicating that the model fits the data adequately. The table of pairs summarizes how many agreements of the number of occurrences of slow and fast exist between the model and the data. There are five times as many agreements as there are disagreements. Our model has shown a significant difference in the proportion of information requests that are processed slow versus fast between the sales regions. It has also allowed us to predict how much more likely a request is processed slowly if it comes from one region over another.

Binary Logistic Regression: Fast/Slow versus Region, Amount

Link Function: Logit

Response Information
Variable   Value   Count
Fast/Slow   Slow   2690   (Event)
             Fast   1388
             Total   4078

Logistic Regression Table

Predictor	Coef	SE Coef	Z	P	Odds Ratio	95% CI Lower	95% CI Upper
Constant	-0.458599	0.0791397	-5.79	0.000			
Region							
East	1.74829	0.100490	17.40	0.000	5.74	4.72	7.00
West	0.624582	0.198423	3.15	0.002	1.87	1.27	2.76
Amount	-0.0000000	0.0000001	-0.32	0.746	1.00	1.00	1.00
Region*Amount							
East	0.0000001	0.0000002	0.63	0.531	1.00	1.00	1.00
West	-0.0000001	0.0000003	-0.46	0.646	1.00	1.00	1.00

Log-Likelihood = -2313.866
Test that all slopes are zero: G = 602.515, DF = 5, P-Value = 0.000

Goodness-of-Fit Tests

Method	Chi-Square	DF	P
Pearson	297.770	330	0.898
Deviance	358.807	330	0.132
Hosmer-Lemeshow	6.960	7	0.433

Measures of Association:
(Between the Response Variable and Predicted Probabilities)

Pairs	Number	Percent	Summary Measures	
Concordant	2387946	64.0	Somers' D	0.39
Discordant	943252	25.3	Goodman-Kruskal Gamma	0.43
Ties	402522	10.8	Kendall's Tau-a	0.17
Total	3733720	100.0		

Figure 6.60 Output From Binary Logistic Regression of Fast/ Slow Process for Information Requests

6.24 The Analyze Checklist

You have spent some time now digging into your data with a variety of statistical tools. The analysis phase of a lean Six Sigma project is usually more complex than the corresponding phase of a traditional Six Sigma project because time data is not described by a well-understood probability distribution. Your data is usually a mixture of fast and slow processes caused by changing priorities for each process step. The relationship between the customer and your company is a dynamic one with potentially conflicting metrics.

Your Analyze phase will consist of three macroscopic parts:

6.24.1 Establish Process Capability

You have validated that you can collect or at least filter reliable data to measure your process accurately and reliably. You can now record your baseline process performance. Collect enough data that everyone, including your customers, will recognize this data as representative of the macroscopic process. Do not attempt to paint a nice picture for management by deleting embarrassing data. Make sure that there is no blame placed on particular people or areas of the business. You will need to characterize your processes in terms of variation (P5-P95 span), central tendency (median), probability distribution, and process efficiency in terms of cycle time.

At the beginning of this step:

- What factors determine the sample size?
- Does your sampling plan take into account seasonal, monthly, weekly, daily, and hourly variation?
- If you are going to take a sample of all transactions, how will you ensure a random and unbiased sample?

At the end of this step:

- Is your data normal, Weibull, log-normal, or a mixture?
- Do you recognize the impact of central tendency and variation on your customer?
- What statistical parameters are you going to use to characterize these indicators? (for example, mean, median, standard deviation, span, and the like.)
- How is your problem statement expressed in terms of these indicators?

Points to remember:

- Define your USL and LSL from the customer's viewpoint and stick to them exclusively throughout your project. Your project improvement is judged with respect to these limits using your operations definition of a defect. They cannot change to make the results look better.
- Spend time on probability plots; you will learn a tremendous amount about parts of your process that are not visible.
- Include all the data. The outliers are probably going to tell you more about your process than the bulk of the rest.
- If your process is capable, then stop the project and return to the Define phase. It may indicate a difference in your understanding of a defect from the customer's viewpoint.

6.24.2 Define Your Performance Objective

The fact that you have been given this project is an indication that the process is not performing as the customer expects it to be. You have defined process indicators as part of your baselining exercise, but now you must establish how much better your process should be to be judged as a success. Customers should not have to collect and analyze data for months to see a difference. The change should be ambitious and visible. If you do not think you can make a big impact, it may be better to spend your efforts on another project. Not all projects are great.

At the beginning of this step:

- What factors will influence the visibility of your improvement?
- What other *critical to quality* (CTQs) might be improved by this project at the same time?
- Do you need to baseline additional CTQs?
- Do you need to do benchmarking to establish your improvement targets?

At the end of this step:

- What will be the target values for the statistical parameters?
- Will your customers feel this improvement? Is it big and is it relevant?
- What was the limiting factor in setting these improvement goals? Do you need to expand the project scope or include other team members?
- Will the business leaders recognize this as a good use of your team's time?

Points to remember:

- When you have multiple customers with conflicting goals, go back to your stakeholder analysis to address any actions required.
- If it appears that you have only one specification limit on a statistical parameter, check the CTQ with other stakeholders. They may impose the missing specification limit.
- Minimize the variation in the process first before working on changing the average. Customers and suppliers are most affected by variation in services.

6.24.3 Identify Multiple Business Processes and Sources of Variation

Transactional business processes are complex and dynamic. You will hear that there is no "typical" customer, service, or process. Dig deeper to map the process in its entirety. When people do not agree, it may be because their view of the process is limited by their speciality—they only use part of the bigger process. There are a lot of individual customer needs that your business is trying to handle. Ask how the organization handles these "exceptions;" they may be the cause of many of your problems. If you have multiple processes, you must flag the transactions accordingly to maintain their distinct qualities for later analysis.

At the beginning of this step:

- What is the business process that is giving you this signal?
- Do you recognize the different ways this process handles different transactions?
- What factors do you think make a difference to the median and span?
- Do you have a diversity of factors from technical, financial, human resource metrics, and organizational communication areas?

At the end of this step:

- Have you uncovered evidence for multiple processes?
- Have you determined what distinguishes one subprocess from another?
- Did you expect different processes and find only one?
- Did you expect one process and find multiple ones?
- Can you recognize the differences in span caused by normal over-loading of a queue versus people changing priorities for personal gain?
- Do you have to rescope the project to include a subset of the multiple processes?

Points to remember:

- You may have to split the project into a few parallel or sequential projects for multiple subprocesses.
- There should be only a few Vital Xs. If you find a large list of incremental factors, you may have missed the biggest ones.
- If you cannot explain the variation in your process, go back to the Measure phase and try again, you may not have identified all the important factors in your data collection plan the first time around. If you still cannot identify the major sources of variation, stop the project here.

7

Improve

7.1 Different Types of Business Processes

Imagine you are a patient seeking medical treatment. If you required immediate attention, you would call 911 and hope *emergency medical service* (EMS) would arrive as soon as possible: fast service measured by low response time is most important. Alternately, your ailment may be a long-term condition requiring a discussion with a doctor, diagnostic tests, and a long-term plan for more extended care. You would phone to book an appointment, choose a time and date that works well within your business calendar and the doctor's other appointments, and hope the doctor could keep the appointments on time: consistent service measured by *variance to want* (VTW) is most important here. Two different problems, fast cycle time and consistent VTW, require a different set of tools during the Improve phase of your project.

The Analyze phase of your project has resulted in an extensive tabulation of your process efficiency, defects, rework, variance to customer wants, cycles of customer wants, and the behavior of these different metrics broken out by different customer segments, service offerings, geographical regions, and other subgroups. Different types of businesses and different segments within a business operate in different ways depending on the relationship between the business and the customer. The type of solutions depend on the relative responsiveness of the business system with respect to the customer demand.

7.1.1 Manufacturing

These business processes generate a limited portfolio of products for their market niche. An individual part or product is not identified with a specific customer until the customer makes a purchase.

These businesses are driven by *return on investment* (ROI) calculations for investment in equipment, material, and labor costs. The historical practice is to group production runs in large batches to minimize total setup time. This results in large amounts of *work in process* (WIP) and finished inventory. The value of WIP can be calculated from its disposal value.

The ideal manufacturing system is poised to manufacture exactly what the customer wants, starting when the customer requests it, by having the order for product "pull" all the components required to manufacture it. In practice, the processes consist of "pull" components to ensure no overproduction and "push" components to expeditiously manufacture what the customers have ordered with no backlog.

Scheduling long runs of a single product increases the average time required to manufacture a product that the customers want, but is not yet available as a finished product. "Pull" systems; kanban, and heijunka are excellent lean tools for addressing these problems. Addressing the customers' needs in terms of product features is usually limited to allowing choices between different options, rather than manufacturing a truly unique product for each customer. Meeting customers' delivery goals requires an understanding of the existing product portfolio.

Execution times for individual steps are similar for different subgroups of product, and long cycle times are influenced by inconsistent delays as a result of changes in supply chain management and production scheduling.

7.1.2 Transactional

These business processes generate a standard product (order, bill, policy), but with individualized features such as address, name, price, and coverage. The elements of uniqueness to each transaction make it difficult to standardize the offering early in the process. These are a bit different from manufacturing processes that have a long lead time because the process can only be started when a customer makes a specific and unique request. The customer initiates the transaction and returns at a later time to complete it.

There is a concept of WIP, but not because of overproduction. It is not possible to build up an inventory of credit checks to be used at a later date. Prepared insurance policies do not get stored in a warehouse because of extra underwriting capacity. It is defects in the form of missing information that cause long rework cycles and influence the total cycle time. Since resources have been expended to create WIP, the cost to the business is the loss of revenue from that transaction, if the transaction is abandoned. There is no disposal value for abandoned transactions.

The primary metric is VTW where the expectation is in the form of a promise to the customer. These processes involve gathering information or conducting research in order to complete a transaction. This could mean assembling various pieces of health information for processing an insurance claim, researching engineering and sourcing requirements for preparing a price quote, conducting employment, reference, and credit checks for preparation of a mortgage. These processes are plagued by a poor understanding of the capabilities of the back office by the front office. Unrealistic commitments are made to the customers, or are poorly tracked as the transaction passes through multiple hands within the organization.

Rework and reprioritization of emergency transactions disrupt the smooth flow of transactions resulting in long delays and variation in total cycle time. The minimization of defects causing rework and delays can increase the capacity of the process and decrease throughput using existing resources.

7.1.3 Walk-in Service

Medical clinics, applications for building permits, and call centers fall under one category owing to the fact that customers must spend time waiting for service. *Work in process* (WIP) includes the customer waiting on hold or sitting in a waiting room before they have received any service at all. Customers may expect to wait a reasonable period of time, but they will abandon you if the wait is too long. In the manufacturing sense this would be considered scrap. No resources have been expended to create WIP, but there is a loss of income when customers leave before they are served.

The business process will deliver an individual, customized response, and it is difficult to predict a standard execution time. The chief problem here is to understand and predict variation in customer volume and make adjustments to production capacity by managing staffing levels, breaks, and extra capacity.

7.2 Different Types of Solutions

The different scenarios discussed above fall in a continuum of philosophies where the goal is to minimize the effect of variation in customer demand on your business processes. The two extremes solutions are "make to order" and "make to inventory". "Make to order" is the perfect lean system. No production of any kind is initiated until an order is received, and no inventory or WIP is required or accumulated. This is not going to be the best solution for a business that is anything other than a high margin, one-off, custom shop. "Make to inventory" is the perfect system of level production. All machines, personnel, and supply lines can be optimized to create the perfectly timed production system for a static product or service offering. Sudden changes in customer demand result in either large amounts of unsold inventory or stockout and back order conditions (Fig. 7.1).

Consider the load on tellers at a bank. The customers choose when they will arrive for service. The demand is created when a customer enters the bank. You must manage this load by predicting customer arrival times and adjust your supply of tellers to match the anticipated demand. This can be as simple as tracking which days your senior customers deposit their social security checks, and schedule an extra part-time staff member on that day. The result is a changing supply that matches the demand with a minimum "inventory" of people waiting in line. This would be an example of a business process that leans towards the "make to order" philosophy.

Now consider the load on a specialized facility for nonemergency surgery. The facility cannot significantly increase or decrease its capacity on an hourly basis like a bank or retail outlet. It manages variation in customer demand by scheduling surgeries and shifting the variation into the "inventory" of waiting patients. This facility, like a dentist's office, must understand the time requirements of different procedures and manage the schedule. This would be an example of a business process that leans toward the "make to inventory" philosophy.

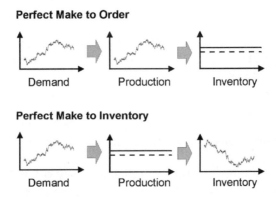

Figure 7.1   Addressing Variation in Customer Demand

All businesses exist somewhere between the two extremes of "make to order" and "make to inventory." Choosing the sweet spot between the two extremes is a constant struggle to optimally match the capabilities of the business with the marketplace.

Toyota has come up with a blend between the two that they call "change to order." The mix of products moving down the entire supply chain is determined in advance using carefully constructed and maintained models of future sales. This appears to be a "make to order" philosophy where the model represents the orders. Toyota does, however, allow some changes to be made to particular vehicles as they are moving though the process. They tightly control how many total vehicles are made using "change to order" versus "make to order." Changing the offering to include a sport package might require changes to engine, transmission, suspension, color scheme, rear view mirror covers, molded body panels, interior upholstery, and seats. Such a major change requires more notice and has more restrictions than changing the type of tires. The optimum balance between the two extremes for your own project will depend on the results of your stakeholder analysis.

7.3 Different Types of Customer Wants

We worked with a business that made customized, computerized monitoring equipment. Each equipment order would have to be defined, engineered, priced, scheduled, tested, and delivered. The cycle time was unpredictable, usually longer than expected and came with an expensive price. There was a choice of twelve different options for a total of 4,096 possible combinations for each unit. Since the total sales were about 1,600 units per year, the business leaders had adopted a "make to order" philosophy.

The general feeling in customer service and the sales department was that customers would order units with a similar group of features. We downloaded the database of orders to examine how many customers ordered customized solutions. We had 1,611 orders with a selection of 12 options for each order. A total of 1,438 orders (89 percent) were ordered with no options whatsoever. We could immediately stratify the business process into a standard line and an option line.

We also wanted to see if there were any correlations between the selection of options. We used Minitab to calculate the correlation between all pairs of options. We copied the results into Excel to create a bubble chart indicating pairs of large correlations (Fig. 7.2). Larger bubbles indicate options that customers tend to order together.

Correlations Between Options

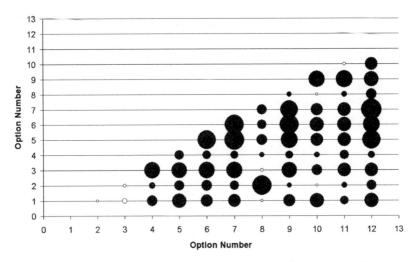

Figure 7.2   Correlations Between Options (Negative Correlations Shown With Open Symbols)

The correlations showed there are some options that are selected in isolation. Option 4 is chosen with option 3, but not usually with any other options. Options 2 and 3 seem exclusive—customers either pick one or the other. Option 2 tends to be ordered without Option 10, while Option 3 tends to be ordered without Option 8. Option 12 tends to be ordered along with options 5, 6, 7, and 9.

Another technique is to investigate if the variables cluster into groups. The Minitab output from the clustering of variables in shown in Fig. 7.3. Near the middle of the dentogram, there is one cluster of options that include 5, 7, 12, 6, and 9. The cluster analysis has identified that this group of options tend to occur together. This is similar to the conclusion found by the correlation analysis in Fig. 7.2.

The analysis from the clustering of variables and the correlations between options indicated that when customers ordered a unit with option 1, they tended not to order other options. This and other similar observations allowed us to divide the product offerings into about six groups, allowing standardization of engineering, some dedicated assembly lines, testing procedures, and other resources.

The final Improve strategy was to allow about 90 percent of the manufacturing and ordering capacity to be directed toward "build to order" for

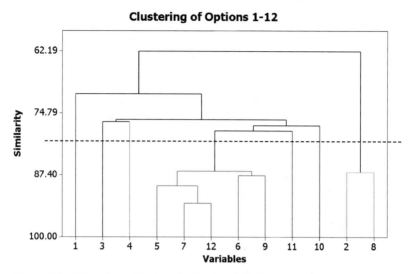

Figure 7.3 Clustering of Options for Product Offering. Dotted Line Indicates 80% Similarity

the "plain vanilla" unit. The regularity of the numbers of "plain vanilla" orders could be predicted using historical data. The remaining clustering of the group of options such as 5, 7, 12, 6, and 9 allowed us to adopt a "change to order" regime to speed cycle time for the remaining six or so groups of product offerings.

7.4 Stratification of Customer and Business Subgroups

The quest for balancing the variation of customer demand requires that you understand the difference between standardized service offerings and more customized solutions. You will never achieve the perfect balance possible by offering only a single product or service.

The analysis of customer needs, delay times, service times, and defects has four different outcomes depending on whether there are differences between subgroups or not, and whether these differences were expected or not. The actions going forward from this analysis are summarized in Fig. 7.4. For example, if there was a different service level agreement profile for internet banking queries versus stock exchange information at a back call center, then the cycle times should reflect this difference. If there was a difference in cycle time for mortgage applications for two different regions served by the same centralized processing center, then your project should address the unexpected causes.

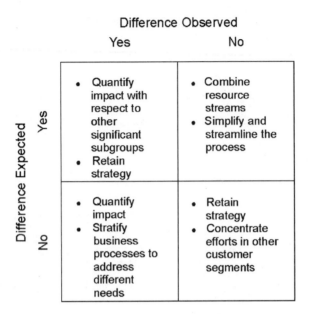

Figure 7.4   Stratify the Business According to Customer Needs

When you stratify your business you could create separate processes for routine and emergency service calls, or seldom and frequent offerings, then you could standardize and streamline the business processes.

7.5 Lessons from Lean and ISO

During the Measure phase, you mapped your process and applied a fairly simple tool to the business process. The *value add/non-value add* (VA/NVA) process-mapping exercise can be applied to any process. In general, extra steps will always add some time to the total. In addition, each process step has a nonzero probability of adding defects resulting in rework or scrap. If the steps can be eliminated without adding substantial risk to the process, then the process cycle time can be improved. It is not always possible to estimate the impact of these improvements, but they generally help.

The first part of an ISO implementation involves process mapping and the preparation of standardized work instructions. During this step alone we have seen ~30 percent improvements. Merely getting the process owners in the room and documenting what they do will improve things by the free exchange of ideas. The other source of improvements is the fear of documenting what turns out to be a pretty poorly designed process. There is a natural tendency for people to improve a process while they are defining and mapping it.

When defects are undetected, they cause enormous problems in the downstream process. The downstream process owners do not feel any ownership, or obligation to correct someone else's errors. Rerouting the transaction back upstream causes large delays and disrupts the upstream processes with "emergency" orders or transactions. Use your process FMEA to identify process tollgates for review and approval. The use of checklists to prevent undetected defects can have a great impact on overall cycle time.

7.6 Kaizen Events

Lean manufacturing uses a number of Japanese terms derived from the *Toyota production system* (TPS) as shown in Fig. 7.5. *Kaizen* refers to the entire process of continuous improvement. A Kaizen event, though, refers to an organized meeting involving the extended group of process owners, stakeholders, and the project champion. You will be asking for ideas for improvement from your team. This may immediately follow the tollgate review when your team presented the results of the Measure and Analyze phases of the project. Respect the enormous amount of experience in your

	Japanese	English	Application
5 Ss	Seiri	Sort	Sort the items and retain only what is needed
	Seiton	Straighten	Organize tools and work space
	Seiso	Shine	Usually the inspection process will expose pre-failure conditions
	Seiketsu	Standardize	Develop systems to montor and maintain the first three Ss
	Shitsuke	Sustain	Maintain a stabilized workplace to encourage continuous improvement
3 Ms	Muda	Waste	Non-value added activities
	Muri	Strain	Overburdening machines or people
	Mura	Irregular	The result of resolving the other two Ms. Production is regular and smooth
	Kaizen	Continuous Improvement	Freqeuntly used to refer to Kaikaku Kaizen or radical change
	Heijunka	Load levelling	Levelling the load of transactions throughout the production system
	Andon	Lamp	Document or lamp board indicating deviation from standard operating procedures
	Kanban	Display card	A visual signal for controlling overproduction

Figure 7.5 Japanese and English Terms in Lean Six Sigma

extended team and give them a structured opportunity for proposing solutions. A lot of great work can come out of these sessions by defining workflow and standardizing work instructions.

There are a number of commonly used strategies for proposing solutions to improve processing cycle time. Most of these techniques have come from the TPS and have Japanese names. The following is not the best translation of the terms you will find in lean manufacturing references, nor is lean the best translation of the philosophy of the TPS.

7.7 Three Ms

When we were investigating the complex feedback systems in extended business processes in Chap. 6, we discussed the "fix that backfires" (Section 6.1). The quick fix of laying off employees to solve the problem of declining revenues caused problems with quality, resulting in declining revenues for the company. The TPS recognizes the relationship of three interconnected components of *Muda* (waste), *Muri* (overburden), and *Mura* (unevenness). If you concentrate on only one of the three components, you run the risk of creating the "fix that backfires" (Fig. 7.6).

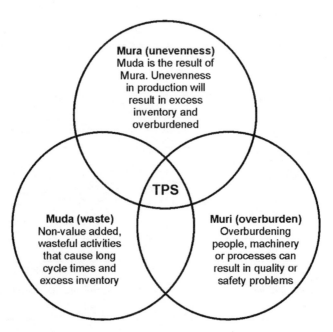

Figure 7.6   The Three Ms

7.7.1 Muda—Waste

There are seven kinds of Muda:

1. Overproduction–Companies often produce work for which there is no immediate customer demand in order to use existing resources. This leads to excess inventory, paperwork, handling, storage, space, interest charges, machinery, defects, people, and overhead. It is often difficult to see this waste as everyone seems busy. This waste tends to take people away from other, more important work.
2. Waiting–Workers may be waiting for parts or instructions, or customers may be waiting for their turn in a queue. Workers typically spend most of their time waiting for one another, which often happens because they have nonaligned objectives.
3. Transportation or conveyance–Poor communication planning can lead to things being moved multiple times and to the wrong people. If roles and responsibilities are not clearly defined, then transactions can be hard to find.
4. Processing–A poorly defined, inefficient process can result in redundant operations, or a customized product where a standardized one is required.
5. Defects–The production of defective services is made at the expense of correct services. The rework causes an interruption in the smooth flow of work.
6. Inventory–Excess WIP is usually a symptom of an inefficient process. Long setup times, breakdowns, misdirected transactions, and production imbalances may hide in long wait times. These problems in the process can be hidden because dissatisfied customers abandon before being served.
7. Motion–This includes movement of people or material. Having everything and everybody at hand reduces the time spent moving order folders, application forms, and other paperwork.

7.7.2 Mura—Unevenness

When we were looking at different ways of handling transactional work, we saw some of the consequences of changing priorities of transactions while they were in process and having to correct defects in transactions by reworking them. These behaviors create enough problems in the system that specialized expediters may be responsible for hand carrying emergency orders through the system. Even though these particular orders are handled quickly, it is done at the expense of every other transaction in the system.

When transactional business processes are designed to be strictly "build to order" with no WIP, then the unevenness of customer demand must be handled by quickly changing the production capacity. The opposite extreme or "build to inventory" also has its problems. Either philosophy will result in feast or famine swings in the production capacity or inventory levels (Fig. 7.1).

7.7.3 Muri—Overburden

Failing to accommodate variation in customer demand can result in swings of over and underproduction. The emphasis for production in times of heavy demand can be made at the expense of overtime, cancelled training, or system upgrades. These factors can result in stress to the workers. The result can be impatience with customers, cancelled transactions, or increase in errors and Muda.

7.8   Five Ss—Minimize Muda

The 5Ss, sometimes expanded to more, are some general techniques that can speed up cycle time of an individual workstation (Fig. 7.5). While these techniques work quite well, they should best be applied to the areas where you have identified problems during the Measure and Analyze phases of your project.

We once had a situation where a manager wanted to apply 5S to all engineering staff in one office. This seemed particularly directed to one engineer's office with enormous piles of documents, reports, and references on all surfaces of his desk, floor, cabinets, and shelves. We conducted a quick survey to see how long it took different engineers to find critical documents. The messy engineer was the fastest in the office. The application of 5S to his office may have sped things up a bit for him, but this was clearly not the biggest problem in the office. Keep this in mind: do not blindly apply these improvements to each process in the business unless you can make a business case for the reason why, and the probable impact of the improvement.

7.9   Heijunka—Minimize Mura and Muri

The first thing you have to do when proposing a solution is to even out or level the production. This is the responsibility of the process owners. They will be the ones who will have to live with the solution and they must be a part of it. This will probably mean that some customers will have to wait for a short, but predictable period of time. Once the production level is constant and predictable, you will be able to apply "pull" systems and balance the production line. When the inventory levels (WIP) have decreased, some of the hidden systemic problems will begin to emerge. This cycle is part of continuous improvement and may spawn a number of other lean Six Sigma projects. Stay focused on the project charter to prevent the problem from becoming too large.

We had one project where chaotic lead times and poor communication had resulted in a very high VTW on product deliverables. One of the biggest problems was that there were inconstant estimates of lead times for components from multiple suppliers. Estimates were usually based on previous experience with similar components. One of the actions for the Improve phase of the project was to contact every external vendor to update or define the cycle times and expedited cycle times for the entire catalog of services and components. This allowed the business to focus on better capacity planning and communication without the variation caused by indeterminate estimates of external supplier performance.

You may encounter some resistance to standardizing work. People may think their value as employees are defined by their unique ability to expedite and execute difficult transactions. They may feel you are not recognizing their creativity and are attempting to make their job into an assembly line. Talk with them and discuss the *Muri*, their overburden, caused by nonstandardized work.

7.10 Define the Queues

One of the most common problems we have encountered in transactional projects is the chaos that is caused by emergency orders, expedited shipments, and special cases. The medical system acknowledges this situation and manages incoming patients by triage. The first step in receiving patients at an emergency facility is to assess their condition and place them into one of three different streams. The downstream processes recognize these three different groups of patients and treat them separately.

The three groups and actions required are:

- Emergent or life threatening–A time delay would be harmful to the patient. Breathing difficulties, cyanosis, cardiac arrest, and spinal injuries are in this group. Stop working on urgent or nonurgent patient immediately to serve this group.

- Urgent–The patient will not get better without medical attention, but the condition will not get much worse with the passage of time. Major fracture, severe pain, burns, persistent nausea or vomiting, and fever are serious, but do not get significantly worse with time. Serve this group in the order in which they arrived, but after patients in the emergent group have been attended to.
- Nonurgent–The condition of the patient may improve without attention, or will not change with immediate attention. Serve this group in the order in which they arrived, but only when all patients in the other two groups have been attended to.

It is common in software development that hundreds of requirements are sorted and managed by using requirements triage. The business requirements or customer needs will provide the necessary grouping for triage in your business process. The way to treat this is to have a clear process for grouping, managing, and updating the priorities of multiple requirements during software development.

7.11 FIFO and Scheduling

For walk in service, the best way to manage the input is *first-in-first-out* (FIFO). This gives the most consistent cycle time for the customers. Examples include teller services at banks, walk-in medical clinics, building applications, and call centers. Allowing customers to be served in any other order will result in one happy customer at the expense of all the others. If there are different service levels for different customer subgroups, then stratify the business by performing triage at the front end and manage each queue using FIFO.

In other transactional services where an expectation time has been promised to the customer, then the order of processing must be calculated using the customer want date/time and the expected cycle time for the transaction. The element of triage has been built into the expectation date by giving a shorter lead time estimate to the customer before the transaction was accepted. This takes more work on the part of managing the workload, but the result should be that individual workers can tell, or are told which of the transactions should be completed before others in their input queue. This also requires that the workload of the people executing the work is known to the front line people making the promises to the customers.

A common example is the scheduling conducted at a dentist's office. This may require the scheduler to get a description of the type of work to be done, the existing schedule of the dentist and hygenisists performing the work, the tasks each one can and cannot perform, and the execution time of each task.

Do not worry about how you are going to achieve this scheduling in your project at this time. You will be assessing a number of different potential solutions once you have defined the different customer segments and service offerings.

7.12 Realistic Cycle Times

You must be able to get a reasonable estimate for cycle time for your sub-processes. This may entail talking to all your suppliers and getting realistic and updated estimates for each service. If you are outsourcing the scanning and indexing of documents, you must be able to plan on reasonable cycle times to coordinate the rest of your internal processes.

It would be a mistake to use the average and standard deviation of historical data to make predictions. It is best to determine realistic cycle times in a different way. We will look at a set of data for the cycle times of a series of transactions with suppliers. We knew from previous analysis that most of the transactions had a reasonably consistent cycle time, but a large number of error prone transactions resulted in a large average. We had already applied some of the techniques in Chap. 6 to segment the transactions into routine and emergency requests. We wished to derive the parameters of the cycle time for the subgroup of routine transactions. The probability plot of cycle time is shown in Fig. 7.7. This probability plot shows the cycle times have a consistent group of short times with a long tail of very long transactions.

Draw a line on the probability plot following the fast process and note the intercepts at the 2.5 and 97.5 percentiles. They are 2.8 and 11.8 hours respectively. The mean for a normally distributed process is half way between these two values.

$$\text{Mean} = \frac{p_{97.5} + p_{2.5}}{2} = \frac{11.8 + 2.8}{2} = 7.3 \text{ hours} \qquad (7.1)$$

This range will define about four standard deviations for a normally distributed process.

$$\text{Stddev} = \frac{p_{97.5} + p_{2.5}}{4} = \frac{11.8 - 2.8}{4} = 2.25 \text{ hours} \qquad (7.2)$$

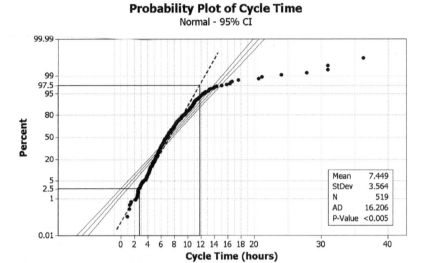

Figure 7.7   Probability Plot of Cycle Time

These estimates show that the variation in the routine process is tighter than it is when all transactions are included in the calculation of standard deviation. When we talked with the suppliers, they agreed that these specifications were within their capabilities and did not require them to drastically alter their processes.

7.13 Stratifying the Business

If you segmented your business into a few different customer streams and designed a business process around each one, you could staff and manage each segment independently, but it would not be the most cost efficient use of internal resources. We need to look at a number of different strategies to find the best way of managing multiple resources with multiple customer segments.

During this section of your Improve phase, you may have a number of different suggestions for configuring your business process to serve different customers with different service levels. This quality of service may also be part of a *service level agreement* (SLA) from either the customer directly or a regulatory body. You might have multiple servers with different skill levels working different shifts, union agreements for minimum number of shifts per worker, and changing customer arrival times requesting different services.

Simple calculations may be made with simple configurations, but there are far too many interacting and conflicting variables to be able to predict and quantify the impact of even a few of them.

There are some general points we can make about the sorts of things to consider. We will not be able to advise you which changes to make, but a few examples can show that the impact of some changes may not be what you expect. When you check in at an airport, there are a number of different ways of configuring multiple servers with multiple types of customers. The bare bones carriers have only one class of passenger, and all customers form a single line. The servers take the next passenger in the line when they become vacant (FIFO). Other airlines give a higher priority to business class than economy class and have two lines. The types of configurations the author has seen are:

- One group of servers who would serve customers alternately from each line. Since there were fewer passengers traveling in business class, their waiting time was less than the economy passengers.
- One dedicated server for business class and four or five servers for the rest of the economy class. The waiting time depended on the relative number of people traveling in business class versus economy.
- One dedicated server for business class who would always serve the next business passenger, unless there were none, then they would take the next economy passenger.

We used ProcessModel to construct a model of a business process to show the implications of changes in strategy for handling two different subgroups of customers with the same group of workers (Fig. 7.8).

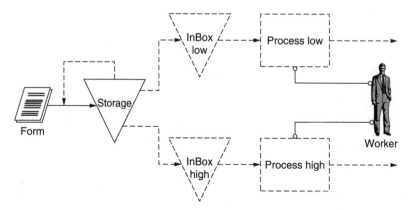

Figure 7.8   Two Queues Handled by a Single Set of Workers

Our business process has two groups of customers with different priorities. They get the same service, but higher priority customers expect to be served quicker than regular customers. About 20 percent of incoming customers are high priority, while the remainder are regular priority. The interarrival time is exponentially distributed with an average interarrival time of 25 seconds. The execution time has a minimum time of 20 seconds with an additional exponential component. It averages 80 seconds. The customers all stand in a single lineup or split into two lineups according to their priority. There are four servers working at about 80 percent of full capacity.

The different improvement ideas include:

- Scenario "A"—All servers serve the customers from a single lineup.
- Scenario "B"—Customers form two lines, high and regular priority, and all servers serve the customers at the front of each lineup at random.
- Scenario "C"—Customers form two lines, high and regular priority, and all servers serve the first customer in the regular lineup unless there is a customer in the high priority lineup.
- Scenario "D"—Customers form two lines, high and regular priority, and one server is dedicated to the lineup of high priority customers while the remainder of the servers assists the regular customers.

The summary of the simulations is shown in Fig. 7.9. The distribution for all cycle times is skewed to high values. The figure summarizes the *P5*, median, and *P95* of the distribution of cycle times. Regardless of the configuration of the servers, the total number of customers of the different types is the same in all scenarios. The difference is how fast or slow the two types of customers receive this service.

Scenario "A" shows when there is no priority given to the high priority customers, the cycle time for the two groups is the same. The median service time is 96 to 98 seconds for high priority and regular customers.

Scenario "B" is interesting in that all servers are serving the customers at the beginning of the lineups without any preference between the lineups. Customers may feel they are all being treated the same way when they are served, but the median cycle time for high priority customers (84 seconds) is less than regular customers (111 seconds) simply because the lineup for high priority customers is shorter than it is for regular customers.

Scenario "C" is an improvement over scenario "B" for the high priority customers. This process has now reached a point where the median cycle time

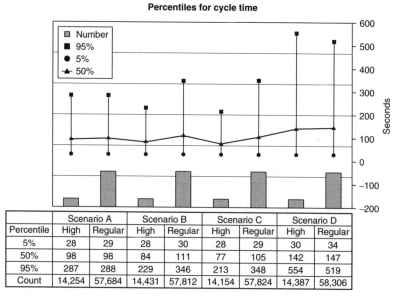

Percentiles for cycle time

Percentile	Scenario A		Scenario B		Scenario C		Scenario D	
	High	Regular	High	Regular	High	Regular	High	Regular
5%	28	29	28	30	28	29	30	34
50%	98	98	84	111	77	105	142	147
95%	287	288	229	346	213	348	554	519
Count	14,254	57,684	14,431	57,812	14,154	57,824	14,387	58,306

Figure 7.9   Percentiles for Cycle Time Using Different Queueing Scenarios and 80% Loading

for the high priority customers is 77 seconds. The median execution time is 75 seconds, leaving a median waiting time of only 2 seconds. Sharing the servers between the subgroups helps even the regular priority customers.

The surprise is scenario "D." This scenario is where we decided to give the high priority customers a dedicated server. Even though 20 percent of customers are high priority, they are getting 25 percent of our servers' resources. Logically, one would expect that having a dedicated server would satisfy the the high priority customers, but the median cycle time for both groups is worse than if they all stood in the same line. In scenarios "A," "B," and "C," the loading of the workers was 80 percent each. In scenario "D" the specialized worker is loaded at 65 percent while the other three workers are loaded at 86 percent each. The specialized line with the extra, dedicated capacity is the worst scenario of the four. The P5-P95 span for scenario "D" is 524 seconds, nearly twice what it is for any of the scenarios where resources are shared between the customer subgroups.

Another scenario was not run, but should be mentioned. Never load your resources much higher than about 90 percent. With the usual variation in customer arrival times, the queue will be empty sometime. Do not make the mistake of assuming that four people at 75 percent load can be replaced by three people at 100 percent load. In this case, any variation in customer arrival time will result in the lineup growing to infinite size.

In order to check the robustness of the scenarios to an increase in customer demand, we ran the models again using 91 percent average loading on the servers. The results are shown in Fig. 7.10.

In scenario "D," three workers are loaded at 97.5 percent while the specialized worker is at 73 percent. In all other scenarios, the resources are equally loaded at 91 percent. Once again the scenario with a dedicated server for one group of customers has the worst performance of the four.

The median cycle time for the high priority customers does not change much with the increased load, but the median and P5-P95 span of the cycle time for the regular customers is about twice what it was when the load was at about 80 percent. The high priority customers in scenarios "B" and "C" do not see much of a change, but the variation in overall customer arrival times has been smoothed out for only one subgroup of customers.

If resources can be shared to serve different subgroups of customers, then the service time for all customers will decrease. This is called load leveling and it is also effective in designing a system that is relatively robust to increases in customer demand. When scenario "B" was increased from 80 to 91 percent load, the median cycle time for high priority customers increased by 14 percent, but increased by about 25 percent when the queue was not leveled between the other servers. This strategy may create problems where the skill level is different between workers. Leveling the load among a group of workers can create personnel problems if some people feel they have to pick up the backlog created by others.

	Scenario A		Scenario B		Scenario C		Scenario D	
Percentile	High	Regular	High	Regular	High	Regular	High	Regular
5%	34	36	33	37	31	37	33	80
50%	144	156	96	187	81	189	177	581
95%	561	580	246	672	220	647	625	1,736
Count	14,693	57,112	14,409	58,000	14,372	57,720	14,243	57,879

Figure 7.10   Percentiles for Cycle Time Using Different Queueing Scenarios and 91% Loading

7.14 Takt Time and Pitch

The average customer interarrival time can be used to calculate the drum beat of transactions as they move through the business. In a business that maintains perfect one piece flow and zero WIP, each business subprocess will process transactions at the same rate. In reality, transactions are split and consolidated, different departments work different shifts with different breaks, some operations are processed in small batches for all kinds of reasons, and individual transactions can have a great range of execution times. It is common that a batch of transactions, a set of drawings, for example, is processed as a unit where the average execution time per drawing reflects the *takt* time. *Pitch* can be used to maintain an average as a target for a group of transactions rather than an individual target for each transaction.

$$\text{Pitch} = \text{takt time} \times \text{number of units} \qquad (7.3)$$

The advantage of using pitch is that the averaging will smooth out the variation in cycle time caused by variation in execution time. The danger of using pitch as a performance metric is that if the time interval is too long, then this will tend to encourage people to process the shortest transactions first, leading to feast and famine cycles. When there are more than a few business subprocesses in a row experiencing feast and famine oscillations, then the span on overall cycle time can increase tremendously. Individual transactions should still be tracked and span calculated on the subset of transactions within the pitch unit. Make sure FIFO is maintained and keep the pitch interval as short as possible. Having pitch as 20 transactions per hour is better than 160 per day.

Workers will tend to use the pitch unit as a batched unit and only pass on groups of transactions as they complete the entire batch. We had one project where a senior auditor would process applications for approvals only once per week. Another business would only make cash calls on Monday. Make sure individual transactions within the pitch unit are handed off as they are completed.

7.15 Kanban in Transactional Processes

When a business process has perfect one piece flow, each business process will generate a "pull" signal to the upstream business process to supply one article for processing. This is extremely difficult to achieve, so a small batch of material is usually "pulled" instead of a single article. This unit, called a *kanban*, is not the same as the size of the pitch unit. When the intermediate supply of an article decreases to a predetermined, minimum amount (safety stock), a kanban signal is sent upstream to generate a batch of intermediate material. The downstream process can keep using the safety stock until the next batch arrives. The calculation of these parameters depends on some economic factors, the size of the batch, stock-out costs, and replenishment cycle time. Variation in inventory delivery cycle time, discounts for large batch forward buying, and changes in consumption rate complicates the calculation (Fig. 7.11). When the inventory of these intermediate articles increases, the ability to economically make quick design changes will be degraded. The effect of variation in customer demand can be smoothed out in manufacturing processes by allowing changes in the level of inventory (Fig. 7.1).

In transactional businesses, the concept of inventory reduction is more about the reduction of rework and batching that is causing WIP and slow cycle times. Production leveling is handled by allowing variation in the number of workers performing a task. In the usual transactional business, the workers are engaged in dozens of activities and process transactions of a similar kind, and process transactions in batches to optimize their own productivity. The previous simulations using queueing models also showed that specialized resources assigned to individual tasks perform much worse than shared resources between the same tasks (Fig. 7.9 and Fig. 7.10). The key to load leveling in transactional processes is to manage the multiple tasks that individual workers are responsible for.

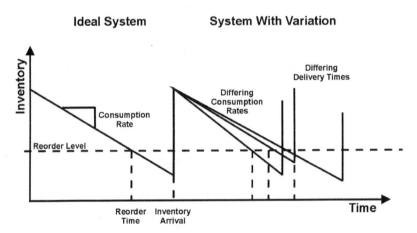

Figure 7.11   Inventory Level Diagram

To illustrate one way of responding to variation in customer demand, we built a model based on a common strategy. Extra workers, managers perhaps, can temporarily handle a rise in customer demand. Supermarkets can have a policy to pull workers off other tasks to work as cashiers when lineups get longer than a certain number of customers. This solution is well suited to short term changes in customer demand. Changing shifts for different numbers of cashiers could be used to handle longer term changes from morning to evening.

The initial situation is that customers are dissatisfied with long waiting times with the present configuration. The four staff members are working at 91 percent loading. As an improvement strategy, we have decided to add an extra part-time staff member who is presently engaged in other duties. This fifth worker is to be called in only when the number of customers waiting is greater than 12—three customers waiting per staff member. The part-time person stops working when the lineup gets below the 12 person limit and they have finished serving the three people waiting in their lineup (Fig. 7.12). The summary statistics are shown in Fig. 7.13.

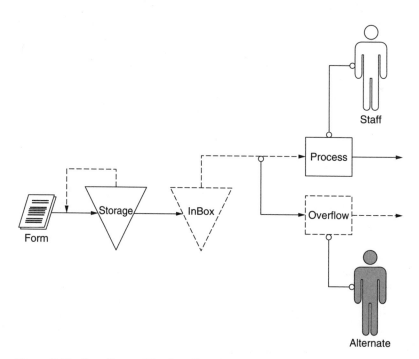

Figure 7.12   Four Servers Plus One Shared Resource for Overflow

Percentile	Four Queues		Four Queues + Overflow	
	Wait Time (seconds)	Number In Lineup	Wait Time (seconds)	Number In Lineup
5%	2	0	2	0
50%	30	4	5	1
95%	150	17	44	6

Figure 7.13   Summary Statistics for Four Servers Plus One Shared Resource for Overflow

The median wait time has dropped from 30 to 5 seconds and the span on number of people waiting in line has decreased by about 67 percent. The average loading of the four regular staff members has come down to 82 percent, while the part-time staff member is spending only 32 percent of his/her time in this new function.

The concept of a kanban card in the transactional context is a signal to add more resources to subprocesses to decrease the waiting time. The previous example showed the effect of a common strategy. More elaborate models could be constructed to investigate the impact of various strategies with differing job description restrictions, different skill levels, minimum shift lengths, and daily changes in customer demand.

7.16   Using DOE with the Model of the Process

Realistic scenarios can quickly become very complex and nonlinear. It may become much more difficult for you to predict the effects of changes and identifying the most important factors in your business process.

We worked on a complex business process that had resisted previous improvement efforts at reducing the variation in cycle time. A number of components of the business process were outsourced, other components were conducted in batches, approvals were done only periodically, different parts of the business operated five, eight hour days per week, and others operated seven, sixteen hour days per week. The team conducted a brainstorming session and generated a large list of ideas with no way of estimating their impact. A process *failure modes and effects analysis* (FMEA) was not of much help in identifying the sources of variation.

We constructed a simulation using ProcessmModel to investigate the consequences of making changes. The process map had over 200 different elements. These 200 elements could be workers, approvals, drawings, quotes, line items, contract terms, locations, partial payments, letters of credit, and so on. Construction of the model itself was a tour-de-force, but even though a good, thorough model had been constructed for the business, we had only changed the form of the problem. We still had no idea where to make an improvement and what effect those improvements could have (Fig. 7.14).

The team assembled all the likely improvement changes. The entire group of stakeholders, nay-sayers included, were asked to consider the impact of making the changes. When we heard objections, they were typically along the lines of, "that depends on...," "that won't work without...," "we tried that and it didn't work...," or, "that will cause this other bad thing to happen." We have learned that whenever we hear those phrases, the process owners are telling us to consider the possibility of multiple interactions between the factors. Whenever you have a situation where there are a large number of factors (there were) or you may have interactions (we had), use *design of experiment* (DOE) to sort out the situation.

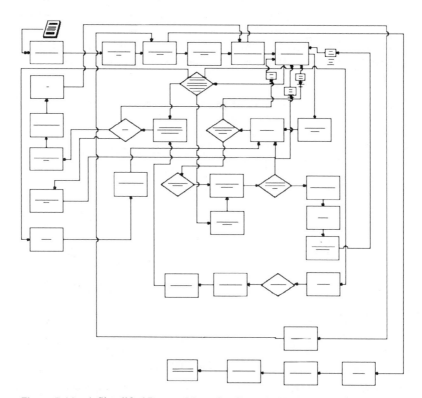

Figure 7.14   A Simplified Process Map of an Extended Business Process

We used the list of factors to conduct an analysis of the model using a 28 run, 26 factor Plackett-Burman screening design. We would alter factors along the lines of adding or moving personnel from one position to another, allowing one process to interrupt another, running extra shifts on some process steps, dividing batches in half, and splitting or joining process flow. The process model was altered according to the experimental design with cycle time as the responding variable. The model runs were performed twice.

The results of the screening design resulted in six factors we wanted to investigate further. We constructed a full factorial, 64 run, experimental design with two replicas and ran it against the model of the process. The Pareto chart of effects is shown in Fig. 7.15.

The largest effects were from the internal1, internal2 subprocesses, and the interaction between them. When we presented this to the team, they pointed out that the internal1 task was a bottleneck and the internal1 and internal2 tasks were performed by the same group of people. Staffing changes in either subprocess would have an effect on the other. One of the subprocesses had been the subject of a previous Six Sigma project, but failed to have the expected impact, because the solution did not recognize the interaction between subprocesses. One group within the team was convinced that the problem was the internal1 subprocess, while another group was convinced that

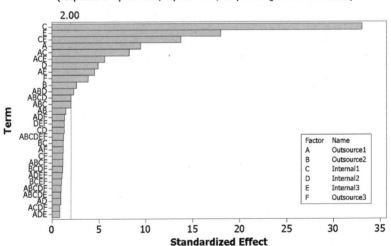

Figure 7.15   Pareto Chart of Standardized Effects for the Process Model

the problem was the internal2 subprocess. They had been attempting to solve the problem by blaming the other group. The model showed reasons why they were both partially correct, and the solution required their cooperation.

7.17 Choosing Between Different Improvement Strategies

The previous sections have summarized some tools for assessing the impact of different improvement strategies against cycle time. The DOE example showed the impact of a large number of strategies against only the single metric of overall cycle time. Each different improvement strategy will make a different impact on each of your other stakeholder groups. Each group may have different requirements for capital spending, time length for implementation, training, hiring, and capability for future growth. In order to select the best strategy we will go back to the *quality function deployments* (QFDs) prepared during the Measure phase (Section 5.8). Recall that the *voice of the business* (VOB) and *voice of the customer* (VOC) surveys were used to identify the key *critical to quality* (CTQs) variables and *critical to process* (CTPs) variables (Fig. 7.16).

We will construct a decision matrix, also known as a Pugh matrix to assess the impact on all *critical to success factors* (CTXs). Without this tool, the stakeholders tend to vote for their own preferred solution based on their own

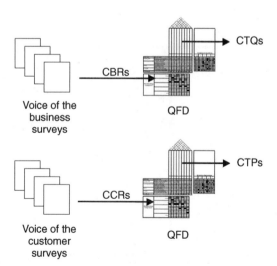

Figure 7.16 Two QFDs Are Used to Identify the CTQs and CTPs for Scoring Potential Improve Solutions

criteria such as speed of implementation, cost, or customer impact. Have the team agree to use the tool, review the validity of the CTX scoring from your Measure phase, and then present the matrix.

Down the left hand column list the CTPs and CTQs with their relative weights as you determined them during your Measure phase. List each improvement strategy across the top of the matrix. Now, in the body of the matrix, score each strategy against all the different CTXs as much better (+3), better (+1), the same (0), worse (−1) and much worse (−3) than the baseline, or existing solution. Ask the team to vote using a show of fingers either up or down. The usual instructions for the Pugh matrix is to rate the improvements as better (+1), worse (−1), or the same (0) as the baseline. An engaged team can easily score potential solutions using the finer discrimination used here (Fig. 7.17). The total score for each potential improvement strategy is the sum of the product of the CTX weights and the interactions. The score for the "improve 1" solution is:

$$(10\times 3)+(5\times 1)+(6\times -1)+(8\times -1)+(3\times -1)=18 \qquad (7.4)$$

The decision matrix can be summarized using a bar graph for presentation.

Your team should now have the documentation required to select the best solution for the stakeholders. You have worked hard to get input from all the stakeholders for potential solutions, and evaluated the impact on CTXs vital to the internal and external customers.

Decision Matrix	Weight	Improve 1	Improve 2	Improve 3	Improve 4	Improve 5
CTQ1	10	3	1	-3		-3
CTQ2	5	1	3	3	-1	
CTQ3	6	-1	-1		3	-3
CTQ4	4		-1	-1	1	-1
CTP1	8	-1	3			3
CTP2	9		-1		-3	1
CTP3	3	-1		3	1	
Total Score		18	30	-10	-7	-19

Figure 7.17   Decision Matrix for Improve Strategies

You may be required to develop an implementation plan for your solution. This plan may be to implement the best solution directly or pilot the solution to work out any unknown issues.

7.18   Establish Improved Capability

During the pilot phase or the initial phase of the implementation plan, you must gather performance data in exactly the same way you did during the Measure phase. The specification limits, operational definitions of defects, and data collection methodology must be the same. This should be relatively easy; the methodology has already been established. You do not have to collect as much data as you did during the Measure phase because the analysis is being only done on the process performance data (Ys), and not the process variables (Xs) required for the identification of drivers of poor performance.

If your project is directed toward reducing the span on VTW, you have done a lot of work to understand the capability of your system to deliver services. If you have stratified the business into two streams for regular and emergency quote generation, you now have a realistic estimate of the cycle time for each subprocess. This must be communicated back to the front line people who are talking, negotiating, and making commitments with your customers.

7.19   Prove the Improve

The changes in median and span for cycle time, or the decrease in span for VTW have been calculated. The sample plan should show that an adequate and unbiased sample of data has been gathered. The Mood's median test can be used to prove the improvement in median cycle time, or median VTW (Fig. 7.18). The Levene's test can be used to prove a decrease in span for cycle time or span on VTW (Fig. 7.19).

Combined probability plots showing the difference in cycle time or VTW are also good to illustrate the nature of the difference in process performance. Figure 7.20 shows the change in overall cycle time before and after the process improvement in a project directed towards increasing cash flow from accounts receivable.

Mood Median Test: AR Cycle Time versus Before/After

```
Mood median test for AR Cycle Time
Chi-Square = 75.03    DF = 1    P = 0.000

                                   Individual 95.0% CIs
Before/After  N<=   N>   Median  Q3-Q1   ------+---------+---------
+---------+
After         625  382   32.00   11.00  (---*
Before        354  491   37.00   25.00                    (---*-------)
                                        ------+---------+---------
+---------+
                                        32.5      35.0     '37.5
40.0

Overall median = 33.00


A 95.0% CI for median(After) - median(Before): (-8.00,-5.00)
```

Figure 7.18   Mood's Median Test Proving a Change in Median Cycle Time for Accounts Receivable Cycle Time

7.20 The Improve Checklist

When you do a thorough job of the Analyze phase, it points you directly to the best areas for the Improve phase. The details of Improve will depend on whether you are focusing on VTW or cycle time. Drop-in medical clinics and call centers operate on cycle time. Quote generation, insurance policy

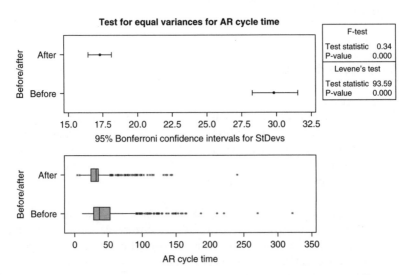

Figure 7.19   Levene's Test Proving a Change in Variation of Accounts Receivable Cycle Time

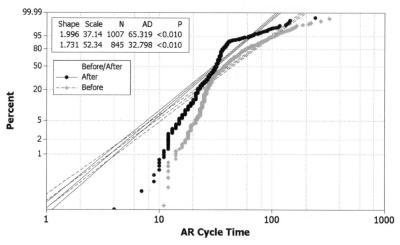

Figure 7.20 Probability Plot for Accounts Receivable Cycle Time Before and After Process Improvement

preparation, loan and mortgage applications operate on VTW. Lean transactional processes should be as linear as possible to maximize the flow of transactions, but this is rare and it is highly probable that there are interactions between the Xs. Keep this in mind as you are proposing solutions.

Your Improve phase will consist of three parts:

7.20.1 Screen Potential Causes

You have chosen whether the customer issue is cycle time or VTW. Part of this phase is understanding the variation in customer demand. This variation can be handled either by leveling the inventory or production. Causes of large span can be tricky to find if individuals are shuffling priorities in their input queues without your knowledge. If there is a reason why the business process should be stratified for different service lines or customer priorities, make sure that the business process behaves that way. If there is a source of variation and you expect it, then things are working fine—anything else is unexpected and should be investigated. When you have identified areas that require improvement, then the application of the 5Ss will usually result in a quicker process. The 3Ms require a balance between reduction of wastes without overburdening the workers or creating unevenness in workload. Focusing on only one of them will create problems with the others.

At the beginning of this step:

- Are you focusing on cycle time or VTW?
- Do you have a short list of Xs?
- Have you checked if the Xs are in the project scope?
- Do you know how the business is stratified?

At the end of this step:

- What other CTQs may be affected?
- What vital Xs have you uncovered?
- Do you know where you will apply 5S and other improvement techniques?
- Do you understand the different customer segments?

Points to remember:

- Check to see if the Xs are real and not just a side effect of something else.
- Resistance can result if people feel that Xs are being used to assign blame.
- The reduction of variation will usually shift the median. Concentrate on the tails of the distributions.

7.20.2 Establish the Relationship Between the Factors (Xs) and the Output (Ys)

By discovering and documenting the relationships between the important drivers of process variation and the output, you should be able to tabulate equations predicting the consequences of changing your process in particular ways. This prediction can be very complex for typical transactional process and will require simulation of the process to both identify key areas for leverage and assessing the magnitude of the effect. Transactional processes show very nonlinear behavior, so consider using DOE to investigate a wide range and combination of strategies.

At the beginning of this step:

- Do you have a list of Xs you wish to investigate numerically?
- Are there restrictions on some of your Xs? Are all the Xs really controllable? Will there be issues regarding changing job descriptions?
- How have you determined a reasonable range of values for your Xs?
- Do you have a well-developed test plan?
- Does the model of your process incorporate realistic restrictions?
- What is your DOE strategy?
- Do you understand the nature of cycle time reduction versus reduction of span on VTW?

- What clusters of customer behavior have you found or not?
- Did you expect the clusters of customer behavior or not?

At the end of this step:

- Do you have a list of all the possible improvement ideas?
- Are defects detected and corrected within the business subprocesses before the transactions move downstream?
- If you are using pitch to determine the pace of work, are you using the smallest time unit possible?
- Do you have prediction equations for all the CTXs for the improvement ideas?

Points to remember:

- Keep the discussion open for process improvement strategies. Encourage people to come up with a variety of ideas, not just the "correct" one.
- When you are constructing a model of the business using simulation, make sure the process owners have plenty of opportunity for input; otherwise they will dismiss the results.
- Include realistic pathways for error correction and rework. They are commonly a major source of variation.
- Overflow workers should not be handling the same workload as regular workers and should not be used to expedite special orders.

7.20.3 Choose the Best Improvement Strategy

It is a mistake to focus on only the one metric of either cycle time or VTW. You will have multiple process outputs for your different process stakeholders (CTQs and CTPs). These have been documented during the Measure phase and will be used here to score different improvement strategies. This phase is where you must ask management and the process owners to make a commitment of time, training, job reallocation, and process changes. These changes are not going to occur just because it will have an impact on a single metric. All the different aspects of the proposed solutions will be considered before making any changes. If the project is piloted, then make sure the pilot is as close to the real process as is possible. In order to prove the improvement consider collecting data from both the new process and old process simultaneously during the pilot phase.

At the beginning of this step:

- Have you involved the team and stakeholders in the process of generating solutions and evaluating results?
- How much variation do you need to eliminate?

- Have you accepted all the suggestions for potential solutions? There should be no surprises when considering the merits of different improvement strategies.

At the end of this step:

- Fully document the new solution.
- Do people at the front end know what is a realistic cycle time?
- Do the process owners know about the consequences of accepting special orders and expediting individual transactions?
- Do you have a plan for the incorporation of inspection and rework for errors?
- Have you documented the improvement and proved the improvement?

Points to remember:

- Do not stop the evaluation of solutions when you find an answer. There will be a large amount of debate by your stakeholders when you present your list of improvement strategies.
- Involve the process owners in the generation of improvement ideas and the evaluation of their effectiveness against all the CTXs.
- Your implementation plan may go in stages. Consider a pilot period with well-planned data collection.

8

Control

8.1 Execution of the Improvement Strategy

Fortune has published a number of articles about why some leaders achieve greatness and why others fail. The lessons that are relevant at this stage in your project are:

- The biggest failure in a leader is not to be without a good strategy, but the failure to execute it. Be focused on execution and make sure the project gets implemented correctly.
- Great leaders create organizations that thrive long after they are gone. The continuing benefit from your process improvement should not depend on you.

The process owners have been participating in the data gathering, analysis, and improvement changes more and more as the project has proceeded. The Control phase is when the *Black Belt* (BB) hands over the new business process to the process owners. If there is resistance to the implementation of the Improve solution, then review your project with the stakeholders and determine what you have missed. Listen to their concerns before you start your Control phase.

We have also seen another reaction at the end of the Improve phase. The process owners may feel that the project has gone on too long and are eager to take on the new process immediately to cease the BB's involvement. Resist the temptation to let the process owners run with the results from the Improve phase. There is still work to be done to make sure that the process changes get implemented. The controls must be in place to prevent the process from sliding back to the way it was before the project began. The lack of a good control plan is the biggest cause of projects failing to maintain the benefits first seen during the Improve phase.

The control plan becomes a part of the new process and continues forever. The process owners may feel this will create extra unnecessary work and could be used against them by management during performance appraisals. Even the word, "control" has negative connotations for many people. Emphasize that a well-executed control plan will put the process in the hands of the process owners and enable them to identify problems before they occur, and define the roles and responsibilities of the process owners and management. The control plan can also incorporate future continuous improvement efforts made by the process owners. The control plan will help them in the long run.

8.2 Change Management and Resistance

When you began your project, you had a goal for each of the *critical to success factors* (CTXs) from the *voice of customer* (VOC) and *voice of business* (VOB) surveys. When you selected the best solution during the latter part of the Improve phase, you made improvements to as many of the CTXs as you could. Projects may not deliver expected results in a number of ways.

The project may have failed to make any improvement in any of the CTXs. This is unlikely, but may indicate that the recommended improvements have not been accepted. We once had a project where the bottleneck in a service process was the two hour turnaround in one subprocess. The three workers could process four requests per day. When the process improvement was piloted, the turnaround time was reduced to 20 minutes and multiple requests could be processed simultaneously. When the improvement process went "live," the turnaround time was about 100 minutes per request. The team heard a lot of objections why the pilot project was not typical and why the results could not be sustained. The simple reason was the workers were worried about job security and found ways of extending the work to fill the day. The project champion and sponsor were called in to ensure that there would be no loss of jobs, and the improvement was implemented with the expected benefit.

It is more common that the project may have made some improvements, but not for all CTXs. Business processes are complex, with multiple interconnected subprocesses. We explicitly considered and modeled the possible interaction of Xs, and considered the impact on multiple Ys using the Pugh matrix as we looked at multiple improvement strategies during the Improve phase. There are real reasons why you cannot expect to increase profit, increase volume, decrease workload, decrease backlog, and decrease cycle time all at the same time. This was the idea of balancing the 3Ms of *Mura* (unevenness), *Muri* (overburden), and *Muda* (waste). Be mindful that solutions may shift the problem to another area and watch for the "fix that backfires." The time delay in seeing the negative effect of some changes can mask the cause of the problem. This is one of the other reasons why you must put control and monitor processes into place and revisit the project periodically until the new process becomes a part of "business as usual."

We had a project where the initial delivery to customer want date was poor. After the process improvement went into place, the customers still had the same comments about poor customer service. The metrics appeared to have improved, until we found that late shipments were being divided into partial shipments so that most of the shipments appeared to be on time. The existing system for monitoring shipment had poor resolution at the transaction level and allowed the process owners to change the problem from late orders to late line items. By monitoring the customer satisfaction, we were alerted to a better definition of process performance than we had used during the Analyze phase.

8.3   Validate the Measurement System for Vital *X*s

One of the results from the Improve phase was a predictive relationship between the vital *X*s and the projects *Y*s. The purpose of the Control phase is to monitor those important process input variables (*X*s) to make sure the process remains stable and consistent. The control plan must ensure the reliability and timeliness of these data in the same manner as was done for the process output variables (*Y*s).

The purpose of the Gage R&R step during the Measure phase was to make sure there was a reliable measurement of the process from the customers' viewpoint. We spent some time making sure that we understood the operational definitions of delivery date, order date, complete shipment, error, and so on. These were the *critical to quality* (CTQs), *critical to process* (CTPs), output signals, or project *Y*s. The tools to validate the measurement system for these measurements are the Gage R&R for continuous and attribute data. A data audit is still a valid tool to make sure that the data is complete and well understood. The *X* data will probably be collected in a different manner than it was during the Measure phase. Many issues arise during the early stages of a project, and one of them is having "quick hit" projects to get more reliable data on *X*s.

When data was being collected for the Analyze phase, a random sample of historical data could suffice for analysis. Data collected for the Control phase, however, must reflect the immediate status of the business process. The ongoing evaluation is designed to detect and alert the process owners when changes in the process have occurred. In an ideal case, deviations from the normal process can be detected before a defect is generated. Corporate dashboards and flash reports are good sources of data during the Measure phase, but these data have usually been averaged and consolidated enough that they are not useful or current enough for the Control phase. The output from the Control phase is to be used at the process owner level, not an upper corporate level. It will not be generalized, but focused on the specific process. The emphasis on individual transactions is even more important here than it was during the Measure phase.

Even though the emphasis is on monitoring the Xs, it is recommended to monitor the process output data (Ys) in case process changes outside the original project scope have an effect on process outputs. Customer expectations can also change over time in the light of technological and competitive changes. Periodically reassess customer expectations and recalculate the process capability.

8.4 Tolerancing

The mathematical relationship between the project outputs (Ys) and process inputs (Xs) has been the goal of your lean Six Sigma project. It allowed you to identify the critical process variables to allow meaningful changes to the process. These relationships also allow you to establish realistic goals for individual process inputs.

This part of the project is where you quantitatively establish the upper and lower tolerance values for a given set of upper and lower specification limits. Imagine that you have established the relationship between the cycle time and the number of errors in a business process (Fig. 8.1).

In this case the *lower specification limit* (LSL) comes from the internal customers who recognize that having the cycle time too fast will result from misallocation of resources. The *upper specification limit* (USL) comes from the external customers who will choose a competitor if the cycle time is exceeded.

When there is no error in the relationship between the Y and the X, the translation of specification limits into tolerance limits is straightforward. Since the regression explains some, but not all the variation in the cycle

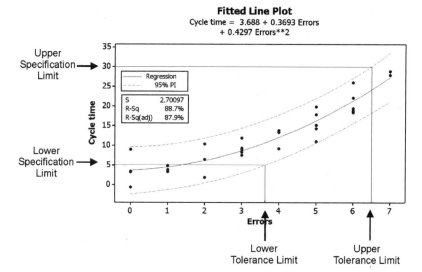

Figure 8.1   The Relationship Between the Number of Errors and Cycle Time

time, the tolerance limits must be narrowed to include the 95 percent confidence interval in the prediction from the regression. The final tolerance limits have been rounded to the nearest integer and set as four and six errors.

8.5   Maintaining Your VTW Goal

Changes in customer demand will require you to constantly monitor and update the models you are using for planning. This can be done using different time scales with different reaction plans. If you were running a walk-in business, then you should monitor the process and resource performance according to a daily, weekly, and quarterly schedule. There will be some overlap between the strategies.

- Daily–For our walk-in medical clinic, the average customer arrival rate changed every 30 minutes or so. Keep track of the number of arrivals for each 30 minute interval during the day, tabulate at the end of the day, and compare with the predicted values. If you have used the improve strategy of having a cross-trained staff member to assist when loading is high, then review how often this staff member was used and when they assisted the regular staff. If the deviations between the actual and predicted customer arrival rates are beginning to change systematically during the day, then consider adding overtime hours at specific times of the day for the regular staff members.

- Weekly–Review the overtime for the week to look for changes in patterns. Consider changing the timing of shifts to address changes. Review overtime hours for the week to see if they are getting excessive.
- Monthly–The original model of arrival times was done using a month of data. It is possible that seasonal changes (winter flu season, summer vacations, children going back to school) may alter the workload from week to week. You will notice these changes if they have persisted from week to week during the month. Consider updating the model using the accumulated data and addressing staff holiday and vacation timing.
- Quarterly–Any changes in the customer demand over a 3 month period are now more in the realm of the business goals and strategy. This may require upgraded training for employees, hiring extra staff, or upgrading facilities.

8.6   Keeping the Process in Control

Segmenting the business process into customer subgroups makes it easier to define a specialized service offering with standard workscope, standard operating procedure, and reproducible risk. The ongoing business process requirements are about maintaining the process improvements and making sure the process continues to deliver the documented performance.

Processes performance can drift when the team has gone on to another initiative and the process owners have not committed to the new process, when unanticipated changes in the process are made, or when changes in other processes have an effect on the process or interest. There are three major techniques used in this phase: mistake proofing, statistical process control, and risk management. A control plan will usually incorporate elements of all three depending on the nature of the data and the process flexibility (Fig. 8.2).

	Example	Control Methodology	Control Element
Standard Process	Completing an application form for a loan	Mistake Proofing	Valid Social Security Number
Changes Possible	Service cycle time	Control Charts	Customer arrival rate
Flexible Process	Sales contract negotiation	FMEA	Risk assessment with risk mitigation plans

Figure 8.2   Three Approaches to Control Depend on the Flexibility of the Business Process

The Control phase also establishes the ongoing plan for preventing errors, process diagnostics, communication, and decision making. A certain number of decisions must be made immediately, while some are long-term and require approval or risk assessment and mitigation. Mistake proofing (*poke yoke*) is the strategy of preventing mistakes from happening. It is implemented once and does not require monitoring. Control charts are constructed to constantly monitor a process and alert the process owner when nonrandom variation from any source has resulted in changes in the process. Risk management is used when each transaction is unique, but a group of experienced process owners have devised a short list of scenarios and action plans to be initiated at tollgates throughout the process to minimize risk.

8.6.1 Mistake Proofing

There are always elements of a business process that are absolutely required to deliver service. During the Measure and Analyze phases these were determined for your process. Any error in these elements caused rework, delay, or abandonment. Some examples are missing purchase order number for collections, missing social security numbers for loans, invalid credit card numbers for loan applications, missing case numbers for medical claims, and incorrect customer contact information for quotes. In these cases it is best to design the business process so that making the mistake is not possible, or at least very difficult.

A common strategy for on-line transactions is to use required fields in application forms. During the Control phase for one project, we used the newly designed web site to apply for a Canadian platinum credit card at a client site. The interface was designed to gather essential information at each step before proceeding to the next. We found a number of opportunities for incorporating mistake proofing:

- The user had to first agree to the terms of the contract before proceeding with the application. This included asserting that the subsequently submitted information would be valid and true. This disclosure would preclude the company from culpability in case of fraud, but would not ensure the accuracy of information provided by the applicant.
- The address information was not restricted to valid street types (street, avenue, close, mews, road, circle, way, and so on).
- In Canada, a valid postal code is ANA-NAN (A–alpha, N–numeric). The leading character cannot be D, F, I, O, Q, W, or Z; the other alpha characters cannot be D, F, I, O, Q, or U. The first three characters correspond to a sortation area and the last three characters are a local

delivery unit. A valid construction may or may not correspond to a valid address. The interface checked ANA-NAN, did not use the invalid character checks or a list of existing postal codes.

- Gross annual income could exceed $100,000,000,000.
- Date of birth had to be DD-MM-YYYY and later than the year 1900, but could be 45-25-1950.
- Whether the city existed in Canada was not checked, but was required. A postal code to address validity check was not performed even at a primitive, provincial level.
- A residential telephone number was required, but would accept European and Asian numbers as residential telephone numbers for Canadian applications. The area code for the resident's telephone number was not checked for consistency with the resident's province.
- Balances could be transferred to other credit card numbers. Credit card numbers reserve the last digit as a check for a valid number. This digit can be used with the Luhn algorithm to check if the credit card number is valid before verifying the number with an external agency. The Luhn algorithm was not used to check for valid numbers.
- The social insurance number was only checked for NNN-NNN-NNN. The last digit of the social insurance number in Canada is used with the Luhn formula to make sure the number is valid in a similar manner to credit card numbers. This property was not used in the interface.

One of the CTQs for the business was accuracy of information during the application process. The Analyze phase of the project had already established that this would cause long cycle times and wasted effort. The Control phase was poorly executed, resulting in a small impact on cycle times.

Many transactional processes involve the passing of documentation from one person to another. The most common strategy we use in these projects is to use checklists that accompany the transaction as it moves through the business process. It is the responsibility of the owners of the process to check for compliance as the transaction moves from subprocess to subprocess.

A project was initiated at a customer site to address billing inaccuracies and long cycle times for preparation of invoices. Pieces of machinery would be accepted 7 days a week for service and repair. The machine shop would check to see if this was a returning customer or not. If this was repeat business, then the machinery was disassembled and inspected. The paperwork that accompanied the piece had an entry for "customer PO number" that could not be completed if the customer's financial representatives were not available. The entry would commonly be completed as "paperwork to follow." The work would be completed and shipped out.

When the job folder arrived at finance for invoicing, the finance department would wait for the "paperwork to follow." It frequently did not. The problems of finding weekend shop supervisors during the week delayed communication. Machinists would rightfully object to being responsible for the financial aspects of the job. They spent most of their time communicating with the direct customer to define the service scope.

The mistake proofing for this project was to print job folders with a space on the outside of the folder for "customer PO number" and the supervisor's name. The finance staff were allowed to reject a job folder for invoicing if it did not have this and a few other pieces of essential information. Finance would frequently give job folders directly back to the supervisors with notes about missing information.

Accurate invoicing was another CTQ for this project. In order to prepare the invoice for the customer, the finance department would frequently have to contact the machinists for clarification on details about the description of the services. When rework was performed, the machinists would not go back to change billable hours on work already performed, so it was usually indirectly billed to the work being conducted when the interruption occurred. Finance people would contact slow paying customers and would again have to talk with the machinists for details about the services they had performed weeks earlier. Even though the existing reporting system had 22 pages of jobs codes for very specific tasks, most of the work was reported using the single code for machining.

The new process was to design a checklist that was started and accompanied each incoming piece (Fig. 8.3). A standard list of inspection tasks was designed for each of the four different business streams. The checklist had check marks for the tasks and spaces for entering diagnostic measurements. When this first phase was complete, the clipboard was placed in an "incoming jobs" rack for the supervisor. He read the inspection report, contacted the customer with the results, determined the workscope and received approval for the work to be done, and entered any essential customer information for billing purposes. The checklist had a list of possible tasks with check boxes for each. He would check the appropriate boxes on the checklist and place the clipboard in the "new jobs" rack for the machinists.

Any available machinist would pick up the next job and proceed to complete the tasks and record on the checklist as to how many hours were spent on each one. When customers contacted the machine shop for job status, anyone

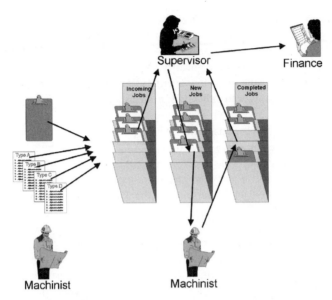

Figure 8.3   Standard Job Scope Definition and Reporting

answering the phone (it was a small shop) could find the right clipboard, read what work had been completed, and tell the customer what work remained to be done. Machinists could also share work among themselves.

When the work was completed, the clipboard was placed in a "completed jobs" rack. The supervisor would confirm that the workscope was completed before handing the paperwork to finance for preparation of the invoice. When we designed the checklist, we defined the job code for each task in the sheet and placed the code number beside the check box on the checklist. Finance would enter the hours against each job code into a small database application to prepare a detailed invoice broken out by task. The finance person worked part time, but when she called into the shop, anyone looking at the rack of "completed jobs" could tell if there was invoicing work to be done.

This project was completed in a very small operation with 5 full time equivalent workers. It was embraced by the workers as making their jobs easier and clarified internal and external communication. The customers received detailed information to decide whether to approve proposed work and finance could prepare well-documented invoices with costs broken out by task.

The project implementation was a very simple one, but we made sure that the workers would use it by having them help with the design of the system.

A few years later, we found the system had been adopted at worldwide locations and was the basis for the design of an electronic job scope definition and tracking system.

Examples of mistake proofing usually make common sense. The most common types of applications we have encountered involve the following.

- Make the transactions more visible. The use of clipboards in the machine shop was simple, but showed everyone in the shop the volume of ongoing work. It is easy to see how many people are waiting in a line up, but difficult to see the average service time. Many electronic queuing systems will permit the process owner to reset the customer number at the beginning of each day. It is a simple matter to hold the last ticket of the day to track the number of customers served throughout the week.
- Use checklists at the tail end of the originating subprocess that must be completed before the work is accepted by the next step in the business subprocess. Error checking and correction can be performed by the correct subprocess owner.
- The workers must own the mistake proofing. The checklist used in the invoice accuracy project empowered the finance staff to refuse job folders that lacked customer purchase orders. This did not require monitoring and escalation by management.

8.6.2   Control Charts and Visual Control Systems

Visual control systems are somewhat related to control charts. These systems are designed to give a constant, visual indication of incoming and ongoing work. In Section 8.6.1, the system of clipboards in the machine shop gave everyone an indication of the amount of incoming work and the amount of completed work awaiting shipment and invoicing. These systems should not be complex. Some elements of mistake proofing can also be incorporated in visual control systems. An example would be that if more than four jobs are completed and awaiting invoicing, then anyone in the machine shop can call the finance person to come in to do a few hours of work. A way of enforcing this would be to limit the number of slots in the rack for completed jobs.

Some business processes will undergo some variation in the input parameters (Xs). The numbers of customers arriving, the cycle time for a particular business process, and the amount of receivables outstanding are some examples of processes that will show a normal amount of variation. You have characterized this variation when you determined the process capability of the improved process. If you have reached this stage of your project, you have established that this new process capability is an acceptable improvement. Control charts have been used for decades in the manufacturing sector and are designed to provide a rapid indication of changes in a process.

The type of control chart depends on whether the lot size of the unit is being tracked and whether the data is discrete or continuous. Variable, or continuous, data may require some form of mathematical transformation before charting. This is not the best solution unless it can be done automatically by software. Asking workers to transform and chart data on an hourly basis will almost guarantee that the control charts will not be maintained.

A common problem in health care facilities is errors made in coding the diagnoses of the patients using the *diagnosis related groups* (DRG) classification. This results in errors in medicare payments and in capacity planning and resource allocation. When a patient arrives at a medical facility, there are a number of different diagnoses recorded in their records:

- Admitting diagnosis–the preliminary reason for the hospitalization. This is done at the time of admission and without much information.
- Principal diagnosis–the condition chiefly responsible for admitting the patient to the hospital for care.
- Primary diagnosis–the diagnosis for which the care and treatment resulted in the greatest consumption of resources.
- Discharge diagnosis–when the patient leaves the hospital, the physician may pick one or many of the possible diagnoses resulting from the accumulation of all data during the patient's hospitalization. The first diagnosis listed may not be the principal or primary diagnosis.

When a patient is discharged, the physician will summarize information on a discharge fact sheet. A coder, trained in medical classification, will assign the most appropriate *international classification of diseases* (ICD)-9-CM code.

Errors in coding the principal diagnoses can be caused by the number of different diagnoses as part of the patient's records or lack of the correct documentation to support the reported code. A project was focused on decreasing errors in coding for two DRGs. The improve plan included training in reporting procedures and escalation in cases of ambiguity. The control plan included the tracking of occurrences of either mismatches between DRGs and ICD-9-CM codes, or DRGs lacking the appropriate documentation before the records were submitted for medicare payment. The total number of patients discharged and the patients discharged with these two DRG codes would vary from week to week. The proportion of cases requiring escalation for clarification was tracked. The control chart could be either a "p chart" to track proportion of escalations, or a "u chart" to track counts of escalations. The results are similar for both charts (Fig. 8.4).

In the early part of the Control phase of the project, we observed a few spikes in escalation rates. When we investigated these errors further we found these spikes occurred on Sundays. The coders did not work on Sundays.

During the week, the coders could resolve ambiguities with the physicians still on duty. The physicians who worked on Sundays were usually not available on Monday when the coding was being done. Any problems with coding would require escalation to the supervisor instead of the physician, overloading the supervisor. This example shows that even though control charts are well designed to detect nonrandom events, an action plan is still required to manage the appropriate reactions and responsibilities.

8.6.3 Risk Management

Risk is defined as the combination of the likelihood of an event occurring combined with the extent of the loss when it does. In the business sense this risk could apply to areas such as the business' reputation, access to capital, access to markets and customers, financial loss, or criminal activity.

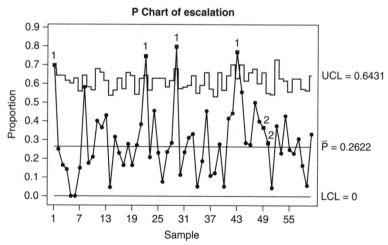

Figure 8.4 Control Chart of Daily DRG Coding Errors for 60 Days

This section is relevant in this phase of your project because you must be able to defend that your control plan can prevent, detect, or transfer risk. This is not to take the place of a corporate audit plan or corporate risk management structure, but the plan must be able to at least document the reporting and decision escalation process you have improved or designed.

Business risk is usually quantified in terms of financial losses. Business risk is currently divided into four or five areas for convenience in management of data and key risk indicators, analytical methodology, regulatory requirements, and risk management methodology.

Credit risk is the risk that the holder of a bond or loan will fail to repay the principal and interest in a timely manner. For financial institutions, this may entail clustering of historical repayments into risk groups. Each group will have its own probability of default, the proportion of the exposure that could be lost, the total amount of financial funds that may be drawn, and the maturity of the exposure. There are practical problems in managing and calculating risk weights for corporations versus financial institutions, emerging markets, or merged firms. A lack of loss data for some groups is common. When reliable data is available, the risk weight functions are calculated and managed using well-established statistical techniques. The overall credit risk can be summarized by applying the risk profiles against the portfolio of outstanding loans. This discipline can be deeply theoretical.

Market risk is the risk associated with an entire class of assets or liabilities. This may be caused by changes in global customer expectations, economic factors, or factors that make an impact on a large portion of the market. Diversification in investments or technology may protect the business from market risk. This may be addressed at the strategic planning level.

Operational risk is harder to define because it is associated with so many different factors. This group includes the risk of loss resulting from a failed or poorly designed internal business process. This could include poor training, poor contract execution, poor approval process, a nonfunctional board of directors, lack of company software licensing, and unclear roles and responsibilities. For example, does the company have a process for entering a legitimate contract or have checks in place to make sure that current capacity is not overcommitted? This is the class of risk most associated with changing and improving a business process and will be addressed in a lean Six Sigma project.

Operational risk will be quantified on an arbitrary scale to identify the portions of the process that will expose the company to the greatest risk in not delivering the service in the time frame that the customer expects. This will be summarized in a risk management or risk mitigation plan.

8.6.4 Operational Risk Control Plan

Refer to the *failure mode and effect analysis* (FMEA) you performed following the process mapping step of your project during the Measure phase. For each step in the process you evaluated the severity, occurrence, and detectability of a failure of the process to deliver the service to the customer in a timely manner. The *risk priority numbers* (RPNs) indicated parts of the process that were prone to problems. In addition to those evaluations, you also documented;

- How often will a failure occur to cause a defect?
- What will be the results of this failure?
- How well will the existing procedures detect this failure mode before it caused a failure?
- Who is responsible for correcting the failure?
- What should they do?
- What are the existing mechanisms for assessing and mitigating risk?
- What are the existing control procedures?

It is common that the entry for "current controls" for the "before" process is "none" during the Measure phase. When a process is improved, the occurrence or detectability should decrease. The severity will remain the same: if and when a failure occurs, then the impact on the customer should not change. The FMEA now has a few more columns to include the risk mitigation plan and the assigned roles and responsibilities (Fig. 8.5). The Pareto chart of the resultant RPNs before and after the process change should show a decrease in operational risk (Fig. 8.6).

When the process team is completing the process FMEA for the improved process, some groups will perform structured loss event and risk scenario modeling. This allows the process owners and stakeholders to propose plausible risk situations to identify how the risk scenarios have been incorporated into the new process. This is done with much more rigor in a *design for Six Sigma* (DFSS) project, but still has some applicability in a *define, measure, analyze, improve, and control* (DMAIC) project.

8.6.5 Quantitative Risk Assessment

Quantitative risk assessment requires historical information on loss frequency and loss size. Statistical analysis of the historical data will reveal the probabilistic properties of the functions, driving factors behind them, and the

Process efficiency
Failure modes and effects analysis (FMEA)

Process or product Name:	Claims scanning and indexing	Prepared by: A. Perkins
Responsible:	A. Hitchcock	FMEA date (Orig)

Process step/part number	Potential failure mode	Potential failure effects	SEV	Potential causes	OCC (Before)	Current controls	DET (Before)	RPN (Before)	OCC (After)	New controls	Person responsible	DET (After)	RPN (After)
Receive and scan documents	Missing documents	Slow review	10	—	7	None	4	280	1	Mistake proof process design with checklist	Scanner	1	10
	Missing FAX pages	Slow review	7	—	6	Check header	3	126	4	Change scan field size for FAXs	FAX operator	1	28
	Duplicate documents	Minimal	2	—	5	None	7	70	3	Sorted documents in case file, manual check by reviewer		3	18
	Mail delivery delay	Scanning delay	6	—	3	Scheduled deliveries	3	54	2	Posted schedule with contact telephone numbers	Mailroom supervisor	3	36
Assign to case file	Illegible	Rework	3	—	7	Operator check	1	21	2	Rescan once, document 'test available copy'	Scanner	1	6
	Assignment to incorrect case file	Confidentiality missing information	4	—	6	None	8	192	1	Mistake proofing process design with checklist		2	8
	Missing date of	Misindex	9	—	6	Operator check	5	270	3	Redesign standard forms, standard stamps	Indexer	2	54
	Workload related delay	Delay	7	—	7	Resource allocation	7	343	5	Daily productivity report (control chart)	Executive	2	70
Review file	Missing physician name	Delay	9	—	4	Operator check	3	108	4	Resubmit and letter to physician	Indexer/physician	3	108
	Missing patient data	Delay	9	—	6	Operator check	5	270	4	Resubmit and letter to physician	Indexer/physician	3	108
	Invalid expense	Wasted resources	2	—	4	Manager check	8	64	4	Checklist of common errors, cancel line item, letter to originator	Manager	5	40
Prepare payment	Case load related delay	Delay	10	—	8	Resource allocation	8	640	3	Daily productivity report (control chart)	Executive	2	60
	Duplicate or missing invoice line items	Incorrect payment	10	—	3	Payment officer check	3	90	3	Sort line items and check for duplicates before submitting for payment to payment officer	Case reviewer	2	60
	Incomplete payment information	Delay	8	—	4	Payment operator check	2	64	4	Redesigned forms with required information	Reviewer	2	64
	Weekly check run delay	Delayed payment	9	—	8	Resource planning	3	216	5	Daily check run	Finance	1	45
	Approval delay	Delay	8	—	5	None	7	280	5	Daily productivity report (control chart)	Supervisor	2	80

Figure 8.5 Process Efficiency FMEA for Medical Claims Processing

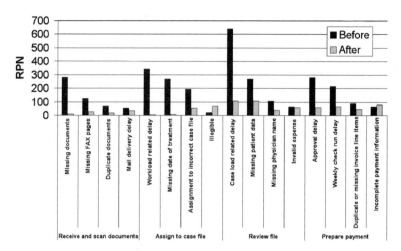

Figure 8.6 Process Efficiency Before and After Process Changes in the Medical Claims Process

subgroups of transactions. If this was the focus of the project as specified during the Define phase, then this information was determined during the appropriate Measure and Analyze phases. We have seen cases where this analysis is done once a year as part of the strategic planning process where capital expenses were projected out for a one year time frame. This analysis was presented to the board of directors, commercial banks, and bond rating agencies.

One control strategy is to evaluate the risk associated with a particular process step, assign a financial value to it, and purchase insurance against the loss. We consider this strategy as one of last resort and should be performed only after all reasonable efforts have been made to minimize and mitigate the risk.

When this technique becomes a part of the control plan for the project, new data must be incorporated and the risk reassessed on a regular basis. The time scale for reassessment should be on the order of the time scale of changes of the significant drivers of variation in risk. When risk capital calculations are calculated as part of the strategic planning process, the projections are typically calculated over a one year time horizon. The uncertainty for projections of data over time increases as the projections get further away from the last data point. This uncertainty can be greatly decreased by periodically updating the projections (Fig. 8.7).

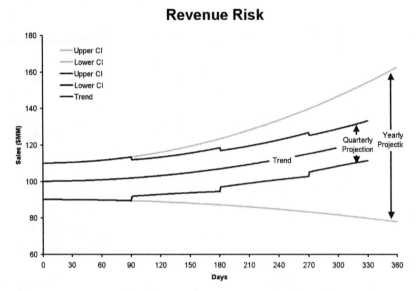

Figure 8.7   Revenue Risk Calculated Yearly and Updated Quarterly

The periodic evaluation of the economic profit of the company can also be considered as part of the control phase of a portfolio of lean Six Sigma projects. Historical analysis will show if a company's past strategy has resulted in a net creation or destruction of value according to the equation:

$$\text{Economic profit} = (\text{ROIC} - \text{WACC}) \times \text{capital}$$
$$\text{ROIC} - \text{Return on invested capital}$$
$$\text{WACC} - \text{Weighted average cost of capital} \qquad (8.1)$$
$$\text{Capital} - \text{Assets relevant to operations}$$

All the elements in this equation are uncertain to some degree. Historical data can be used to project capital requirements and economic profit for a future time frame, and these projections should be updated periodically.

Another scenario requiring periodic risk reassessment is in large project management. The initial estimates of costs and time frame are poor. As the project progresses, these projections change and the significant drivers of indeterminacy change. A periodic reassessment is required to update the risk profile of the project or project portfolio.

We had a project to assess the financial risk associated with fraud losses. The profile of fraud loss would change almost monthly due to changes in technology and detection. The new business process was to reevaluate the expected financial exposure on a more frequent basis. This allowed the company to purchase the best short-term fraud insurance. The time profile of the fraud risk showed a similar pattern to that shown in Fig. 8.7.

8.7   The Audit Plan for Project Close Out

There will always be portions of the business process that will not change exactly as you had planned. It would be unrealistic to expect the customer needs and the business process to become perfectly stable after the process has been altered. We typically find an initially large change in performance when the project nears the Control phase and before it rebounds to a stable state. When the project has closed out, the strategic focus of the business may also have shifted to another area. Unless the Control phase includes a periodic check on the benefits, there is a real possibility that the benefits will diminish to zero after a few months.

The first audit of financial benefits is done at the Control phase after the process improvements have been made. The initial estimate for the financial benefit of the project done during the Define phase is just that, an estimate. At the time it was sufficiently accurate for the *Master Black Belt* (MBB) and the executive team to balance the relative merits of different projects when they were deciding how to prioritize the portfolio of projects. Now that the process improvement is in place, a much more realistic quantification is required. Use the financial flow chart to categorize the benefits (Fig. 8.8).

The rules established during the Define phase will apply: if the benefits are greater than about $50,000, then the team's financial representative should approve the calculation. If the project has resulted in reduced costs, then an estimate of the costs of running the new process can be made. If the benefit is estimated as a percentage of sales and sales are expected to increase, then specify that the benefit is a decreased cost of sales per unit rather than an overall increase in revenue.

One year after project closure, a second audit is required of projects that have a large financial benefit. This prevents the benefits from being claimed without a realistic tracking process to ensure continued benefit. An example would be a new process that has succeeded in clearing out a large amount of *work in progress* (WIP), but does not continue to keep the WIP at a low level. Once the process has been in place for one year, then the new process

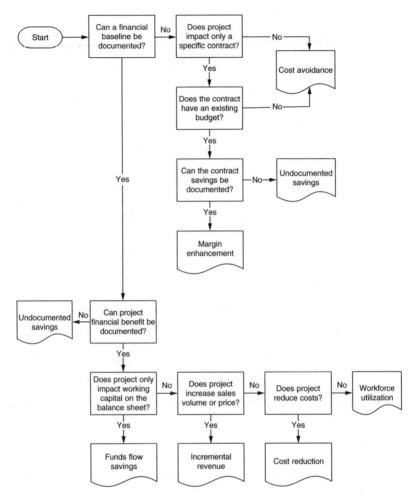

Figure 8.8   Financial Benefit Flowchart

is considered as the regular way of doing business. You cannot claim the same benefit year after year.

The categories of uncertain financial benefit are the undocumented savings, incremental revenue, workforce utilization, and cost avoidance groups. This may be because the company has only a low volume of the services addressed in the project. Once a year has passed, then the past year of sales or services in that area can be audited, tabulated, and classified as incremental revenue. Historical trends in failing to meet service level agreements, price concessions for late delivery, or overtime expenses for a given service level can be compared from year to year.

8.8 The Control Checklist

Many lean Six Sigma projects and entire implementations fail when there is a lack of discipline in the Control phase. It feels like the heroes have been recognized and the awards given out. The design of the control plan seems like a superfluous phase of the project and the new process is poorly transferred to the process owners. When the stakeholder analysis, communication plan, or resistance plan are poorly maintained, many projects will cease at this point.

A well-designed control plan will define when the process has gone out of control, and who is responsible for doing what to correct it. Not only does the control plan define when workers can go to management for corrective action, but also defines when management should stay out of the process and not micromanage the process and workers.

Your Control phase will consist of three macroscopic parts:

8.8.1 Validate the Measurement System for the Vital Xs

This step has the same flow as the step that considered the reliability of the Y data during your Measure phase. The main difference is that the ongoing data collection for the Control phase must be fairly automatic and timely.

At the beginning of this step:

- What are the vital Xs as determined from the Analyze phase?
- How are the vital Xs measured now?
- What did you notice about the Xs while you were conducting your Measure and Analyze phases?
- Does the production environment differ from the test environment in your Measure?

At the end of this step:

- Is the data collection process stable?
- How large is the error on the Xs?
- Does the ongoing data collection process need improving?

Points to remember:

- An ongoing audit plan for validity of data collection may be required. In the manufacturing environment, measurement equipment is calibrated periodically and the procedure is part of the control plan.
- Have a finance representative participate and approve the calculations for projects with large financial benefits.

- New systems may interact with your ongoing data collection. Check to maintain the definitions and measurement systems. Adding new departments or sales offices may require a process owner to conduct a data audit for the new input.

8.8.2 Establish Operating Tolerances on the *X*s.

The equations relating the customer specifications on the *Y*s need to be translated into operating tolerances on the *X*s. You have done a lot of work in establishing these relationships. The responsibility of the process owners is to stay within the tolerances on the *X*s. You must provide these for the process owners.

At the beginning of this step:

- What are the specification limits on the process *Y*s?
- What are the relationships between the *X*s and the *Y*s?

At the end of this step:

- Can you combine any of the tolerances on the *X*s to cover more than one *Y*?
- Have the tolerances been incorporated into the standard operating procedures?

Points to remember:

- The tolerances on the *X*s are the final result of extensive analysis. They are not subject to changes by process owners after the fact. There is a tendency to slowly relax the tolerances when process owners find it is running well.
- When the process is operating efficiently, there is a tendency for monitoring to be considered as a redundant and wasteful activity.
- Calibration should be incorporated into the control plan and become part of the new process. It should not be an external QA function.

8.8.3 Implement Process Control and Risk Management

When effort is not put into monitoring a process, other business initiatives and entropy in general will result in the process drifting over time. The best control plan is easy to maintain and provides everyone with the data to detect problems before they become too severe and the boundaries to stay away from a process that is operating well.

At the beginning of this step:

- What is the level of acceptance for your control plan?
- Does the control plan require a minimum of work on the part of the process owners?
- Have you made maximum use of mistake proofing?
- Have you incorporated elements of visual control?

At the end of this step:

- Do the process owners understand their responsibility in maintaining the control plan?
- Have you documented the process improvement?
- Did your FMEA identify points in your process where you should incorporate tollgate reviews?
- Have the decision points in the FMEA been incorporated into the standard operating procedures?
- Have you considered using quantitative risk assessment and management?
- Do you have an audit plan if you need to finalize financial benefits after one year?

Points to remember:

- The best solution for managing risk in a process is to eliminate it by automating the process step or using mistake proofing in some way. Make sure that the mistake proofing ensures that an error-free transaction proceeds to the next process step.
- Operational risk is based on historical data and scenarios. Use risk scenario modeling to plan for anticipated risks.
- Operational risk should be periodically reassessed as part of the control plan.

9

Sustain

9.1 Results

During the Recognize phase, the executive defined the criteria for selecting and prioritizing projects directed at customer needs. Where the Recognize phase was about alignment, the Sustain phase is now about accountability. This phase is not as linear as the phases of a *Define, Measure, Analyze, Improve, and Control* (DMAIC) project. The three main components are:

1. Passing metrics from the DMAIC projects to the Sustain phase at the completion of each project
2. Integration of lean Six Sigma into business systems
3. Ongoing assessment, support, and development of the lean Six Sigma program

9.2 Maintaining the Six Sigma Program

These functions are chiefly the responsibility of the lean Six Sigma program custodian. This person will be the VP-lean Six Sigma, VP-Process Excellence, or a combined VP/*Master Black Belt* (MBB) role. This responsibility may also lie with a lean Six Sigma or Process Excellence steering committee (Fig. 9.1).

The ongoing assessment, support, and development functions of the Sustain phase will seem familiar and appear like components of a DMAIC project. At the conclusion of a DMAIC project, data are passed to the program custodian by the MBB. These data are used in the Sustain phase in an ongoing effort to assess the health of the program and the effectiveness of the direction set by the executive during the Recognize phase. The important difference is that during a DMAIC project, data was being gathered and analyzed for a specific business process. In the Sustain phase the data being gathered and analyzed is about the lean Six Sigma program itself.

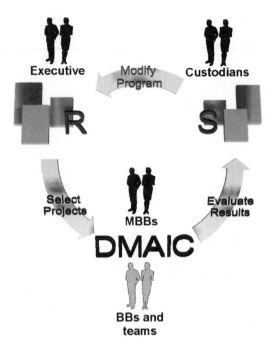

Figure 9.1   Roles and Responsibilities for R-DMAIC-S Phases

In keeping with the DMAIC methodology at the project level, the portfolio of completed projects is periodically treated as a representative sample of lean Six Sigma projects at the program level.

This accountability component of the lean Six Sigma program has three stakeholder groups.

1. The progress on the project Ys are related to the criteria used to select the projects to begin with. These criteria were first defined by the executive and documented in the corporate *Quality Function Deployment* (QFD) section of the Strategic Planning QFD. The MBB helped select which of these metrics is applied to individual projects during the Define phases by focusing on the "internal" QFD. Now the reports on those metrics go back up through the Strategic Planning QFD from the DMAIC teams to the lean Six Sigma custodian for evaluation. The program custodians are responsible for collating and tracking this data and summarizing it to the rest of the executive.

2. The process owners are somewhat interested in the improvements in the Ys, but their focus should be on the factors they are responsible for, the vital Xs. The vital Xs have become part of the control plan.

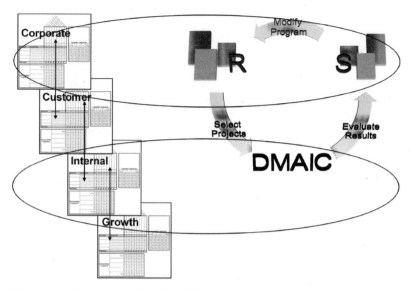

Figure 9.2   The Strategic Planning QFD and the R-DMAIC-S Cycle

The information on the Xs does not flow back up through the customer or corporate QFDs and is not shown in Fig. 9.2.

3. The MBBs have guided project teams to ensure a valid application of the DMAIC methodology. The teams improved a business process and transferred it back to the process owners during the Control phase. There are a lot of measurements used to monitor and diagnose problems associated with the growth and health of the lean Six Sigma program. The MBBs are responsible for collating and tracking this data and summarizing it to the lean Six Sigma custodians.

The major portion of the Sustain phase is involved in the upper arrow in Fig. 9.1. The upper arrow also addresses components of the growth QFD of the Strategic Planning QFD.

9.3 Ongoing Financial Benefits

There were two calculations of the financial benefits done during each lean Six Sigma project. The first was an estimate done during the Define phase, during project selection and approval, or early in the Measure phase. The financial benefits flowchart (Fig. 8.8) was used to categorize the benefit. The second calculation was done when the project was completed and may be updated during the final project audit performed 1 year after close out. These results were reported and tracked by the MBBs.

The program custodian must include the cost of training, recruiting, project implementation, and program administration before reporting the financial impact of the program at the corporate level.

Figure 9.3 shows the results from a mature lean Six Sigma implementation. The numbers have been scaled, so the weighted average program cost is 100 units per year. During the first partial year of implementation in 1999, there were a large number of lessons learned during the Sustain phase. The Recognize phase of the program was substantially redesigned in late 1999 to focus on projects with quantifiable financial benefit. Some early projects also had to wait for the audit performed 1 year after close-out to be included in the calculation of the impact of the program. In late 2000, the Sustain phase was more tightly integrated into the Recognize phase for use in annual strategic planning. The benefits from the program continued to grow thereafter as the program was better integrated into the hiring, succession planning, strategic planning, and program administration processes from year to year.

Let us suppose that the savings project benefits were not what was expected, the customers had not noticed an impact, the *Black Belts* (BBs) were not applying or learning the tools, or the lean Six Sigma program was not identifying and retaining the future leaders of the company. In these cases, the

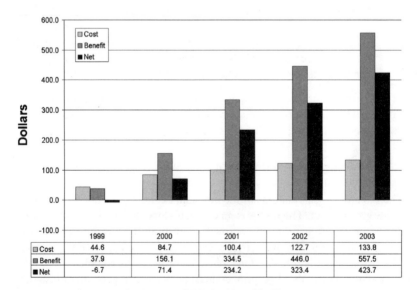

	1999	2000	2001	2002	2003
Cost	44.6	84.7	100.4	122.7	133.8
Benefit	37.9	156.1	334.5	446.0	557.5
Net	-6.7	71.4	234.2	323.4	423.7

Figure 9.3 Return on Investment (ROI) for Lean Six Sigma

evaluation of the lean Six Sigma program has a *variance to want* (VTW) flavor to it. The source of the mismatch could come from one of the two sources:

1. The Define Phase
   - What effect did an individual project have against the initial targets set out in the Define phase for the project?
   - Were the tools used correctly?
   - Were the expectations and metrics clearly understood?
   - Was the champion engaged?
2. The Recognize Phase
   - What effect did the portfolio of lean Six Sigma projects have against the targets set out in the Recognize phase for the lean Six Sigma program?
   - Is the training material relevant?
   - Are the projects identified and well scoped?
   - Are the BB candidates suitable for leadership development?
   - Are the upper business metrics and strategic goals guiding the project selection?

Your lean Six Sigma custodian must be able to assess the driving factors behind the success or failure of the program. Some of the factors have been listed above, but you must evaluate your own program in the same way you evaluate any large, new business initiative.

9.4 Reporting and Tracking Projects

We have intentionally kept this discussion in the Sustain phase because it is one of the more important factors in sustaining the success of subsequent waves of projects and lean Six Sigma resources.

The difference between executing DMAIC projects and managing the lean Six Sigma program creates quite different requirements for reporting depending on whether you are a BB with a single project, an MBB with a portfolio of active projects, or a lean Six Sigma custodian with a few dozen dedicated resources and hundreds of archived and active projects.

Individual project teams must have the tools required to plan and manage their own projects. We find that the project teams do a fairly good job of updating the communication plan, planning agendas, publishing minutes of meetings, preparing project charters, and preparing presentations for tollgate reviews. We believe that the company can have the teams manage their own projects manually. A project binder is commonly used to organize the various components of the project and is maintained by the BB. The program custodians should design checklists for the BBs based on the summaries at the end of the Chaps. 4, 5, 6, 7, and 8.

We encourage keeping hard copies of important documents, such as approvals at tollgate reviews and minutes of meetings. If these components become integrated into an automated system for project tracking, it encourages champions and sponsors to become less involved in the projects. MBBs, champions, and sponsors can then lose track of the undocumented interpersonal issues that can ruin the best projects and the team can feel the project has become less important.

The lean Six Sigma custodian's and MBB's roles require a dedicated project tracking system. Managing a set of projects and managing lean Six Sigma resources is much more complex than managing a single project. The MBBs need to track the projects as they are progressing and analyze the historical project data after the fact. This data should be kept in standardized form for all completed projects. When this database of completed projects grows, it becomes a great resource for the program. It can be used to search for examples of projects to use in customizing the training material, tracking the systematic poor use of some tools, can be a cause to modify or augment the training program, or MBBs can track milestones of individual projects to predict reasonable time lines for new projects.

A database that can keep track of potentially hundreds of completed projects, each with their own different types of financial benefits, cycle times for completed phases, and team members and roles, requires a well-designed system with ties to the company's financial reporting and *human resource* (HR) systems. The details of the implementation of the tracking system will depend on the existing financial reporting, project management, document management, HR, and process management systems. It is vital that the database have a search facility based on any aspect of the projects.

We have seen too many companies underestimate the effort required in sustaining the lean Six Sigma effort by relying on Excel spreadsheets and collections of PowerPoint presentations. The continuing value of the company's program depends on accumulating the lessons learned, time lines, tool usage, financial expectations, internal resources, and process knowledge. It is common for BBs to underestimate the amount of work required to properly document their projects, and to wait until the end of the project to assemble the important information. This lack of planning will result in the creation of a repository of enormously valuable information that is impossible for the

MBBs to use in program planning or the BBs to use as a source of lessons learned from previous projects. As the implementation becomes more mature there is more and more potential for projects to overlap, or potentially undo the benefits from completed projects.

9.5 Lean Six Sigma Corporate Dashboards

Corporate dashboards, or automated data gathering and reporting systems are becoming more common. If lean Six Sigma is going to be a priority for the company, and an effective tool to transform the culture, then its targets and results should be visible and presented on an equal footing with other more traditional, financially-based measures of company performance. We have been at a number of implementations where important customer based metrics were tracked instantly and displayed prominently. Examples have been weekly summary of variance to want for delivery, span on call length and hold time, and number of orders processed hourly. These metrics were incorporated into existing executive dashboards systems equipped with some historical and analytical ability. At one company, daily lean Six Sigma metrics were displayed as a news program type "crawl" at the bottom of the internal company website. Individual metrics could be clicked and analyzed. Another company showed the first page of the daily corporate Six Sigma dashboard as a screensaver on all company computers.

In the early part of a lean Six Sigma implementation, other metrics should be tracked and reported in a similar manner. Examples include the number of people trained and certified at the GB, BB, and MBB levels. Active projects at different DMAIC stages should be broken out and summarized by *critical to quality* (CTQs) and *critical to process* (CTPs). Estimated impact on CTQs and CTPs should also be tracked. The custodians of the program should do *voice of customer* (VOC) and *voice of business* (VOB) surveys to determine what the stakeholders expect of the program, independent of what they expect from the projects.

The summary metrics of the health and status of the program should be integrated into quarterly reporting. It is especially important to update and summarize the financial benefits every 3 months. The financial portion of lean Six Sigma going to external financial analysts must be reliable and traceable. This reporting should include how the projects and CTQs are lined up with corporate goals as a part of strategic planning. When an implementation is mature and lean Six Sigma has become part of the corporate

DNA (the "way we do business") then it becomes less of a specific line item in the strategic plan and more of an enabler of strategy.

9.6 BB Assessment and Certification

The original goals of the program were to identify, train, and promote the future leaders of the company and to infuse the corporate culture with a data-based decision-making business improvement process. There are a number of HR functions that are responsible for assessing and maintaining this goal.

Responsibilities and expectations for project team members should be clear and measurable. The characteristics of a successful implementation from the HR viewpoint include:

- High potential candidates are identified.
- High potential candidates are applying for BB positions.
- Retention of BB candidates is high.
- Retention of certified BBs is high.
- Ex-BBs move into operational roles.
- Ex-BBs in operational roles continue to be retained and promoted.
- Candidates and graduates learn the soft skills of project and change management in addition to the hard skills of data gathering and analysis.

This book has concentrated on the statistical aspects of lean Six Sigma, but it is a mistake to assume that statistical wizardry defines a good BB. The qualities of an effective leader can only be assessed in the light of a portfolio of completed projects. Effective projects may be statistically simple, but require a great deal of change management and conflict resolution. The best BBs are the ones who can combine the quality of the solution with the depth of its acceptance by the process owners (Equation 9.1).

$$\text{Effective} = \text{quality of solution} \times \text{acceptance} \qquad (9.1)$$

The "quality of solution" term can be assessed by one-on-one interviews with the MBB or combined with a written examination. The written examination is based on the body of knowledge and the competency model for the role (Fig. 9.4). The "acceptance" portion of the equation can be assessed with performance reviews based on the completed projects.

Black Belts (BBs) must be in their role for 18 to 24 months before an accurate assessment can be made about their effectiveness. They should have completed over half a dozen projects by this time. Most large implementations accept

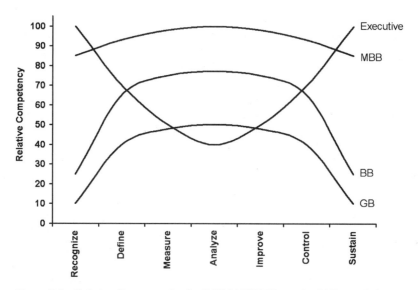

Figure 9.4   Relative Competencies for R-DMAIC-S Phases for Different Roles

BBs from outside the company, but still require them to take the internal training and recertify using the company's criteria. This recertification is less commonly required of MBBs. The reason new MBBs take the internal training is that they are expected to teach it as part of their new position.

Most of the learning about how to best apply lean Six Sigma takes place during project execution. It would be ideal to have everyone gain the same experience from being a part of dozens of projects, but this is not practical. Project teams can learn a lot by interacting with other groups during periodic off-site meetings. Having a database of completed projects goes a long way in helping the *Green Belts* (GBs) see what worked in the past.

Black Belts (BBs) and MBBs must be much better then the GBs in tool usage and project management. They will never really learn this until they have to teach it to others. *Master Black Belts* (MBBs) must spend some of their time training subsequent waves of GBs and BBs. It is also expected that individual sections of the training material should be presented by competent BBs. This involvement of the BBs in training should increase as they become more comfortable. This helps drive the culture of team-based problem solving and leadership development. *Master Black Belts* (MBBs) can make notes about problems with the training material, present examples of tool usage during training, and use the participants' projects during working sessions.

A minimum amount of time conducting training is part of the role definition of the BBs and MBBs. Good reviews from the learners are a part of their assessment for certification.

Remember that your lean Six Sigma program is directed to training your own future leaders. This deeply specific direction precludes using external agencies for certification. *Black Belt* (BB) certification is awarded by the program custodian with recommendations from the MBBs. *Human resource* (HR) departments must be able to discuss certification criteria when BBs are hired by other companies.

9.7 Maintain a Body of Knowledge

There are different roles played by the project teams, BBs, MBBs, project sponsors, champions, and executive. The lean Six Sigma custodian must define and maintain the body of knowledge required of each role. We have seen highly structured matrices of skills and tools required for each designation extended into GB, *Yellow Belt* (YB), and *White Belt* (WB). The logistics of defining up to eight different levels of skills is truly daunting, let alone having to maintain training material, certification requirements, and transition requirements for each level. On top of this, keeping HR records up-to-date for all the participants is extremely difficult.

The purpose of the lean Six Sigma program is to identify, train, and evaluate future leaders of the company. There is no reason that a potential future leader should learn or not learn about different aspects of lean Six Sigma. Successful implementations maintain at most, four levels of competency on the same set of tools depending on peoples' roles.

Our model is to maintain one set of training material for all levels and choose the amount of detail covered during training depending on the audience. It is best that everyone have at least some knowledge of all tools in the lean Six Sigma toolkit. The competency at each level is different, but the material remains the same. Figure 9.4 shows the relative competencies expected for the different roles. The general differences between the executive and the "Belts" reflects the emphasis of the roles in the *Recognize-Define, Measure, Analyze, Improve, and Control-Sustain* (R-DMAIC-S) framework (Fig. 9.1). The competencies have a similar profile for each of the "Belts" with higher requirements from GB to BB to MBB. The chief difference between the MBBs and the BBs is that the MBBs must communicate with the executive in the Recognize and Sustain phases (Fig. 9.4).

We are unaware of many projects that use all the tools in the toolkit. *Black Belts* (BBs) must be able to select the most appropriate tool at each stage for each project depending on the circumstances. This can be done by consulting the MBB in cases of ambiguity. Even if BBs and GBs never use some tools for years after training, they will eventually have to be aware of their limitations and their applicability.

New tools can always be added to the appropriate sections in the R-DMAIC-S framework. An example is the addition of nonparametric statistical tools into the Analyze phase. These were never part of the original program because the manufacturing applications usually generated normally distributed data. When Six Sigma began to be applied to transactional projects, new tools were added and became part of the body of knowledge.

The program custodian should conduct a periodic assessment of the effectiveness of the lean Six Sigma program by talking with the customers to see if they are really feeling the impact of the program. It is regrettably common that Six Sigma is used as a vehicle for cutting costs with little or no impact on the customers. Piet van Abeelen, the VP Quality of Six Sigma at GE, would travel worldwide and to all GE businesses to assess the effectiveness of the Six Sigma program at the local, customer level. He constantly concentrated on real results at the customer level and rigor in DMAIC project execution.

9.8 Communication Planning

An important component of a healthy lean Six Sigma implementation is the celebration of successful work by the teams. This is the responsibility of the Six Sigma custodian using data provided by the MBBs.

Every company has some form of internal communication. This could be a periodic newsletter, the internal website homepage, web casts, or periodic company meetings and announcements. All successful lean Six Sigma implementations have some kind of communication plan that encourages participation, builds momentum, and "buzz" for the program.

We have done a number of projects centered about HR issues and find that the success factors for the program includes visibility of the participants. Internal visibility to senior executive is very important for team members and external visibility of the program to customers is important for the company. Internal communication can be a little more open with respect to details about the project, the participants, and financial benefits. External communication can be in the form of conference presentations and other publications.

Early in a lean Six Sigma implementation, there will be a certain amount of teaching the language of Six Sigma to the company. Publishing a photograph of the team members along with their project's defect definition and problem statement is sufficient to show that the company has made a commitment to solving some problems that are probably already well known to the people in the company. They should be able to appreciate the impact the problems have on the customers.

As the projects proceed and the implementation matures, showing the major milestones of the DMAIC process is more important for driving the culture than the short descriptions of the solutions to the problems. These short descriptions are suitable for external communication where the goal is to show how lean Six Sigma is being used. The continuing emphasis of internal communication is to show the details of the DMAIC process and the people involved. Changes in corporate culture can be initiated at the corporate level, but real change in lean Six Sigma comes from the empowerment of the improvement teams. Internal communication should also identify when new BBs are selected and sent for training and when BBs are moved out of their roles and placed in more senior positions.

A health care provider implemented Six Sigma a few years ago. We followed the external and internal communication plan carefully. After the announcement of the program to the board of directors, the communication plan centered around educating the general employees. In the first year of the implementation, the monthly internal newsletter carried a progression of articles on the program at the "internal" and then customer levels of Fig. 9.2. As the implementation matured, the communication plan shifted more to an external focus at the customer and corporate levels (Fig. 9.5).

There were a remarkable number of good points made in this communication plan. These include showing:

- The value of good data
- The power of unbiased analysis
- The occasional surprise during the projects
- Aspects of change management and team dynamics
- The value of cross-functional teams
- The projects' impact on the customers
- Definitions of the roles and responsibilities of team members
- The stages of the DMAIC process

Date	Communication
Oct 2002	Integration of Six Sigma into strategic planning
Jan 2003	Short note in report of strategic planning session to shareholders. "GE did it and is helping us to save a lot of money"
Jan 2003	Internal announcements to the staff as part of report of strategic planning session. More details on the process with a reference to the identified candidates
May 2003	Newsletter - defines Six Sigma as a new way of doing business and applying successful management techniques to health care
Aug 2003	Newsletter - defines Six Sigma roles and names of positions. Outlines how this is part of the existing process improvement and leadership development programs
Sep 2003	Newsletter - first stage of training done and projects are defined (titles and brief problem statement). Names and photographs of black belts and project teams
Sep 2003	Newsletter - defines roles of black belts as change agents with ties to leadership programs
Oct 2003	Newsletter - status report of an operational project nearing completion, some details on the process mapping exercise and lessons learned. Emphasis on cross functional communication and team learning
Oct 2003	Newsletter - status report to finance project being split into one on collections and one on accounts receivable
Nov 2003	Newsletter - interviews with team members talking about the unbiased use of data to find surprising answers and identifying a very few vital Xs driving a majority of the variation in the processes
Dec 2003	Newsletter - interviews with team members for another project with similar results to the Nov 2003 newsletter. An emphasis on the DMAIC methodology
Feb 2004	Newsletter - status report and preliminary results from four projects
Apr 2004	Newsletter - first review of impact on Ys. Quotes from customers noticing the impact on their CTQs
Jun 2004	Newsletter - second wave of five projects announced. Also points out that two projects are replicating the success of the first wave in a different department
Aug 2004	External publication - interview with CEO appears talking about the lessons learned in implementing the program. Impact on the customers and shareholders, selecting projects, changing culture. Published in a medical publication
Nov 2004	External publication - summary of one individual project in detail with Six Sigma tools and language. Published in a Six Sigma publication
Jan 2005	Conference talks by Six Sigma custodian at quality conferences

Figure 9.5   Health Care Facility Six Sigma Communication Time Line

An important point is that all projects were presented as they progressed. The executive were committed to the program and pushed following the DMAIC methodology over just making changes. All projects were followed regardless of their eventual impact. Everyone in the organization knew about the projects as they progressed. If the teams had to talk with someone new, the new members already had an idea what was expected of them, and how the project was supported by the upper management.

There is great value in creating a lean Six Sigma intranet site for communication within and between the BBs and project teams. Lists of contacts, examples of spreadsheets, templates for reporting, and project check sheets can be posted for general use. It is straightforward to sequence the stages of a generic DMAIC project, link each stage with the best tools, and have examples and explanations for each step. The result can be an interactive *lean Six Sigma coach*. Once built, the interactive project map could even be used to track milestones of project execution.

9.9   New Projects

A very important issue continuing the success of lean Six Sigma is the maintenance of a new project *hopper*. Some of the sources of projects were discussed in the Recognize phase. The development of these project ideas may require some assistance by the MBB in the Define phase of individual projects. The program custodian and MBBs must continue to define the program's direction in the Sustain portion by compiling project ideas from various sources such as:

- Data integrity and reliability issues from active projects
- Problems discovered during project execution, but "out-of-scope"
- "Quick hit" projects generated during the Measure phase that became too large during subsequent project definition and scoping
- Managers in the field who received GB training or a program overview and are aware of the resources available to help them address problems
- Executive level lean Six Sigma steering committee
- Financial analysis of warranty returns, customer attrition, scrap, and so forth.
- Customer survey results
- Competitive benchmarking

9.10   The Sustain Checklist

Lean Six Sigma is not something that will transform a company in a few short months. It will take many years to change the culture of the company to one, which is customer centered, data driven, and process focused. The Sustain phase is less of a phase and more of a continuous process of assessment and modification of the lean Six Sigma implementation.

The Sustain phase will consist of three macroscopic parts:

9.10.1 Report and Evaluate the Six Sigma Projects

Metrics from DMAIC projects have been gathered and passed to the program custodian when each project was completed. This data is used to assess the overall impact of the program on the goals set out by the executive during the Recognize phase. Some of these will be financial metrics such as cost savings, incremental revenue, or margin enhancement, while others will be related to customer CTQs such as delivery to want, or warranty returns. These metrics will be defined by the program custodian and will probably be part of a lean Six Sigma project tracking system.

At the beginning of this step:

- Has the financial representative to the DMAIC teams approved the calculation and classification of financial benefits?
- Have the MBBs reviewed the final presentation of the project?
- Have the process owners been trained in the new process and accepted the changes?
- Do the process owners understand that they now own the new process and Control plan?

At the end of this step:

- Have project benefits been entered into the project tracking system in a manner that allows for roll up summaries to be generated and circulated?
- Has the project documentation been stored and indexed for the program custodian or MBB?

Points to remember:

- This is the very last step taken by the DMAIC team. Make sure the process owners have accepted any future obligations, such as the retention of records, to allow an audit of financial benefit after 1 year.
- The data gathered on completed projects must be comprehensive enough to be able to analyze a portfolio of projects for assessing and improving the program.
- Examples of relevant data are tollgate time lines, the involvement of team members, metric improvement, and tool usage.

9.10.2 Integrate Lean Six Sigma into Business Systems

The lean Six Sigma program is a major business initiative no different in scope and importance than any other. The program must be integrated into the existing reporting and management systems. The long-term goals of lean Six Sigma require the program to be integrated with HR, finance, and operational reporting systems. The ability to analyze a portfolio of completed projects is required to maintain and modify the *body of knowledge*, certification requirements and status, and corporate communication planning processes.

At the beginning of this step:

- Have you determined all lines of communication and reporting within the company?
- What analytic ability is required for the management of the program?

At the end of this step:

- Is the lean Six Sigma program visible at the corporate level?
- Are the lean Six Sigma program elements visible where they become integrated with the other business systems?

Points to remember:

- The evaluation and reporting systems are to be used to identify gaps in the general execution of DMAIC projects, not to conduct individual performance appraisals.

9.10.3 Identify New Opportunities

Every business is different in the sense that every business will have different types of problems that are the most critical elements of a strategic plan. Each implementation of lean Six Sigma will be slightly different owing to these local differences. As the company integrates BBs into operational roles the goals of the implementation will continue to change.

At the beginning of this step:

- Have you identified gaps between the objectives of the program established during the Recognize phase and the assessment of closed-out projects?
- Are these gaps owing to an unclear direction of the program, poor execution of DMAIC projects, or inability to select the projects with the highest potential impact?

At the end of this step:

- Do you have an ongoing process to generate and evaluate new projects?

Points to Remember:

If the lean Six Sigma program is not delivering what the stakeholders want, then:

1. Recognize there is a problem with the program.
2. Define what the stakeholders want out of the program by talking with them.
3. Measure what the program is not delivering by looking at past projects.
4. Analyze the key drivers of the process to deliver the results you want.
5. Improve the program by changing it.
6. Control the program.
7. Sustain the development of the new program and be aware of the need to evaluate it.

EPILOGUE

The kernel of Six Sigma will never go out of date: use data to make business decisions with a customer focus. Never lose sight of that vision as Six Sigma continues to grow and as you continue to learn.

> All of our big, Company-wide initiatives have led us down serendipitous paths, and Six Sigma has proved no exception. It has, in addition to its other benefits, now become the language of leadership. It is a reasonable guess that the next CEO of this Company, decades down the road, is probably a Six Sigma Black Belt or Master Black Belt somewhere in GE right now, or on the verge of being offered—as all our early-career (3–5 years) top 20% performers will be—a two-to-three-year Black Belt assignment. The generic nature of a Black Belt assignment, in addition to its rigorous process discipline and relentless customer focus, makes Six Sigma the perfect training for growing 21st century GE leadership.
>
> John F. Welch, Jr., *Chairman of the Board and CEO*
> Jeffrey R. Immelt, *President and Chairman-Elect*
> Dennis D. Dammerman, *Vice Chairman of the Board and Executive Officer*
> Robert C. Wright, *Vice Chairman of the Board and Executive Officer*
> *GE Annual Report–2000.* February 9, 2001

Quantitative Risk Assessment

A.1   The Effect of Uncertainty on Estimation

There is always some element of uncertainty in the estimation of the total cycle time, financial benefits, or cost for any extended project. The *program evaluation review technique* (PERT) is one example of the use of estimation of cycle time in project management.

Any business process you consider will consist of not just one, but many successive steps to achieve an outcome. We are going to learn a technique developed for the construction of mechanical devices and apply it to the estimation of cycle time for business processes and financial risk for pricing. In the broader scope, this technique falls under the discipline of tolerance analysis.

A.2   Monte Carlo Simulation

The Monte Carlo technique is simple in concept, but computationally intensive. The power of desktop computers has enabled a numerical tool that was once restricted to the research domain to be applied in a variety of general applications. This can be illustrated by a simple example.

We have a simple business process that consists of two steps; people wait in a lineup then fill out an application form. Each step has a one-in-six chance of failing. What are the chances that someone entering will fill out an application form correctly?

One approach to solving the problem would be to mathematically express the probability of each occurrence and combining the probabilities correctly. This problem is slightly complicated by the fact that the events are not independent: when someone leaves the lineup, they do not attempt to fill out an application form at all.

The Monte Carlo technique is another method of tackling the problem. Have (simulated) people enter the lineup, choose them one at a time, and have them throw a pair of dice. If they throw "doubles," they leave, if not they stay. The people who stay are asked to throw the pair of dice again. If they throw "doubles," then this is counted as an incorrect application, if not, it is counted as a successful application. Repeat this process for a very large number of (simulated) people, 1000 to 10,000 would suffice. Count the total number of (simulated) people who stood in line and calculate the proportion of correct applications.

When the properties of the random elements of a process are well characterized, the process can easily be simulated tens or hundreds of thousands of times.

Crystal Ball is a set of Monte Carlo routines from Decisioneering (www.decisioneering.com) that are installed over the top of an existing Microsoft Excel installation. The installation adds a new toolbar and can be used on any existing Excel spreadsheet. More advanced users can incorporate Crystal Ball commands in macros. We will construct a few examples to show how it can be used in lean Six Sigma projects. The first example is to investigate the consequences of a small error rate on both the external customers and internal business stakeholders for insurance policy underwriting. The second example is a more complex one where we will use *time value of money* (TVM) to investigate the financial and business impact of delays in a medical claims processing center.

A.3   Cycle Time for Insurance Policy Underwriting

The term for the entire process of insuring a risk is called *underwriting*. The usual practice is to have a large number of insurance agents contacting customers while the preparation of the policies themselves is undertaken by a centralized underwriting team who evaluate the applications, price the risk and write the policies. A high-level process for preparing a property insurance policy is shown in Fig. A.1.

The process begins when an insurance agent makes a request for under-writing to the centralized insurance underwriting facility. The request is first distributed (DISTRIBUTE) to the underwriting team assigned to that agent for review, risk evaluation, and classification (RATE). The underwriting team hands the policy to the pricing team, who calculate the schedule for paying the premiums (PRICE). The policy is finally passed to the writing team who assemble the final document (WRITE) before it is electronically sent to the originating agent for presentation to the customer (RETURN).

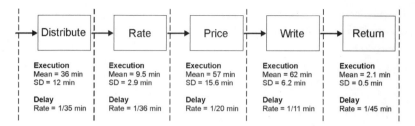

Figure A.1   Preparing a Property Insurance Policy

We have gathered time stamped data for a large sample of policies and determined the parameters shown in Fig. A.1. Each execution step shows a normally distributed execution time with the indicated means and standard deviations. Preceding each step is a delay step which follows an exponential distribution with the indicated rates, where the rates are the reciprocals of the individual average delay times (for example, an average delay of 10 minutes equals a rate of 0.1 per minute).

Even though the distribution of time for each of the steps in the underwriting process are known, it is very complex to derive a mathematical expression for the total cycle time. We will use the Monte Carlo technique to generate thousands of random samples from each distribution and then manipulate them to derive the quantities of interest.

The process is simple in concept:

1. Generate a random delay time for the distribution step from an exponential distribution with a rate of 1/35 minutes.
2. Generate a random execution time for the distribution step using a normal distribution with a mean of 36 minutes and a standard deviation of 12 minutes.
3. Generate a random delay time for the rating step from an exponential distribution with a rate of 1/36 minutes.
4. Generate a random execution time for the rating step using a normal distribution with a mean of 9.5 minutes and a standard deviation of 2.9 minutes.
5. Continue generating random delay times and random execution times for each of the three remaining steps in the process.
6. Repeat steps 1 to 5 a large number of times (~10,000).

From this large set of data, quantities such as total execution time, total delay time, total cycle time, and span on total cycle time can be easily calculated.

A sensitivity analysis will be conducted to determine which of the multiple steps makes the greatest contribution to the variation in total cycle time.

A.3.1   Construct the Model

Parameters from the insurance underwriting process in Fig. A.1 were entered in column B in an Excel spreadsheet as shown in Fig. A.2. The file was saved as "Underwriting.xls." Crystal Ball was started and the spreadsheet opened.

The auditing tool shows the calculation of the average total execution time in cell C23, the average total delay time in cell D23, the average cycle times for each step in cells E4, E8, E12, E16, E20, and the average total cycle time in cell E23. Enter the formulae to perform the summations as shown.

We are going to follow a number of steps to determine the answers to our questions about the underwriting process.

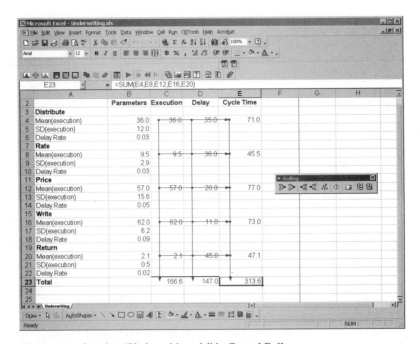

Figure A.2   Opening "Underwriting.xls" in Crystal Ball

A.3.2   Define Assumptions

First enter the distribution information for the execution time of the distribution step (Fig. A.3).

1. Select cell C4.
2. Select "Cell>Define Assumption..." from the drop down menu or click the "Define Assumptions" button on extreme left of the Crystal Ball toolbar.
3. Double click the "Normal" distribution graph.
4. Enter the information from Fig. A.1 into the input boxes for "Assumption Name," "Mean," and "Std Dev" as shown in Fig. A.3. Make sure that you have entered cell references such as "=B4," by typing them. Clicking the cells in the spreadsheet does not work.
5. Press the "Enter" button, otherwise the distribution data will not be updated.
6. Press the "OK" button. The dialog boxes will disappear and cell C4 will now be colored, showing it has been defined as an assumption cell.

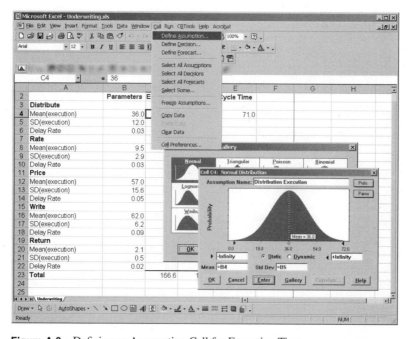

Figure A.3   Defining an Assumption Cell for Execution Time

Now enter the distribution information for the delay time for the distribution step (Fig. A.4).

1. Select cell D4.
2. Select "Cell>Define Assumptions..." from the drop down menu or click the "Define Assumptions" button on extreme left of the Crystal Ball toolbar.
3. Double click the "Exponential" distribution graph.
4. Enter the information from Fig. A.1 into the input boxes for "Assumption Name" and "Rate" as shown in Fig. A.4. Make sure that you have entered cell references such as "=B6," by typing them.
5. Press the "Enter" button to update the distribution parameters.
6. Press the "OK" button. The dialog boxes will disappear and cell D5 will now be colored, showing it has been defined as an assumption cell.

Continue the process of entering the parameters for the execution and delay time distributions for each one of the steps in the process, Rate, Price, Write, and Return from the data in Fig. A.5.

The distribution for all the execution steps is normal and the distribution for all the delay steps is exponential.

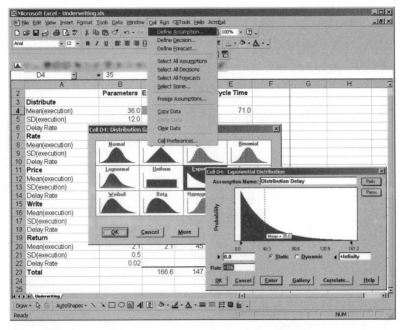

Figure A.4 Defining an Assumption Cell for Delay Time. All Crystal Ball Toolbar Icons Have Been Blurred Except for the "Define Assumption" Icon

Cell	Assumption Name	Distribution Type	Mean/Rate	StdDev
C4	Distribution Execution	Normal	=B4	=B5
C8	Rating Execution	Normal	=B8	=B9
C12	Pricing Execution	Normal	=B12	=B13
C16	Writing Execution	Normal	=B16	=B17
C20	Returning Execution	Normal	=B20	=B21
D4	Distribution Delay	Exponential	=B6	-
D8	Rating Delay	Exponential	=B10	-
D12	Pricing Delay	Exponential	=B14	-
D16	Writing Delay	Exponential	=B18	-
D20	Returning Delay	Exponential	=B22	-

Figure A.5 Data for Assumptions Cells in the Underwriting.xls Example

A.3.3 Define Forecast Cells

The spreadsheet must now be set up to gather data and statistics for predictions based on the assumptions and characteristics of each process step. The total cycle time for process step is defined in column E as C+D, the addition of the execution time and delay time (Fig. A.6).

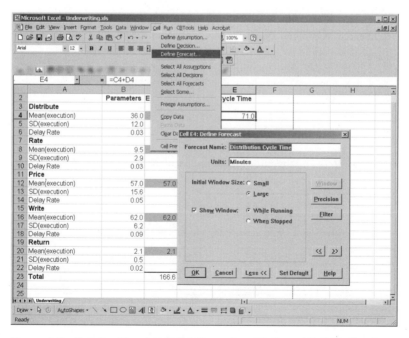

Figure A.6 Entering Forecast Cell Information. All Crystal Ball Toolbar Icons Have Been Blurred Except for the "Define Forecast" Icon

1. Select cell E4.
2. Select "Cell>Define Forecast..." from the drop down menu or click the "Define Forecast" button on the Crystal Ball toolbar.
3. Enter the information for "Forecast Name" and "Units" as shown in Fig. A.6.
4. Press "OK."
5. Cell E4 will change to a different color than the assumptions cells, indicating it as a forecast cell.

Repeat the above procedure to define "Forecast" cells for each total cycle time in cells E8, E12, E16, and E20, including the grand total cycle time in cell E23 using the data in Fig. A.7.

When the simulation begins to sample and resample the distributions for the execution step and delay step, the additions in column E will be repeated each time (Fig. A.6).

A.3.4   Set Simulation Parameters

We are now ready to run the simulation to gather the data you need. Most of the default settings are adequate for this example. We will only change the number of samples we are going to take (Fig. A.8).

1. Select "Run>Run Preferences..." from the drop down menu, or click the "Run Preferences" icon in the Crystal Ball toolbar.
2. Click "Trials" and enter 10,000 for the number of trials.
3. Click "Options" and check the "Sensitivity Analysis" box.
4. Click "OK."

A.3.5   Run the Simulation

You have entered all the information on distributions, forecast cells, and run parameters. It is time to find out how your process will behave.

Cell	Formula	Forecast Name	Units
E4	=C4+D4	Distribution Cycle Time	Minutes
E8	=C8+D8	Rating Cycle Time	Minutes
E12	=C12+D12	Pricing Cycle Time	Minutes
E16	=C16+D16	Writing Cycle Time	Minutes
E20	=C20+D20	Returning Cycle Time	Minutes
E23	=SUM(E4:E22)	Total Cycle Time	Minutes

Figure A.7   Data for Forecast Cells in the Underwriting.xls Example

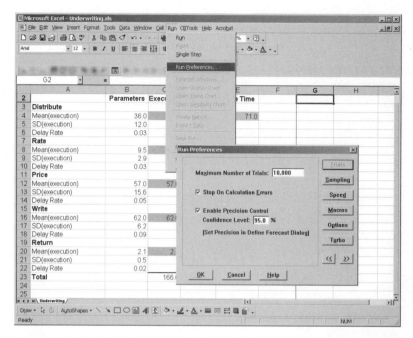

Figure A.8   Setting Run Parameters. All Crystal Ball Toolbar Icons Have Been Blurred Except for the "Run Preferences" Icon

1. Select "Run>Run" from the top of the drop down menu shown in Fig. A.8. You may also click the icon in the Crystal Ball toolbar to the right of the "Run Preferences" button. It looks like a VCR style "play" button.
2. Watch the screen as the histograms for the forecast cells are updated while the simulation repeatedly samples the probability distributions and summarizes the time forecast calculations (Fig. A.9).

A.3.6   Evaluate the Results

Crystal Ball will accumulate summary statistics for each forecast window. These can be viewed by selecting the forecast window of interest and then selecting "View>Statistics." Since there is a random component to the simulation, your simulation will have similar, but not identical results to the simulation shown here (Fig. A.10).

A.3.7   Sensitivity Analysis

We now have a summary of the range of execution times that the predicted process would take. We would also like to determine which of the process steps has the greatest impact on the total cycle time.

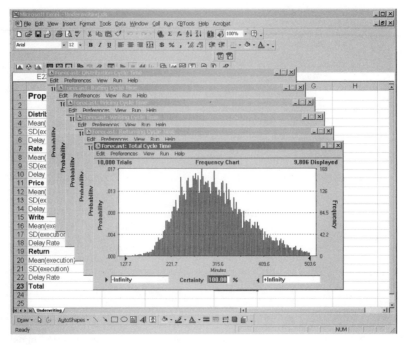

Figure A.9   Forecast Windows

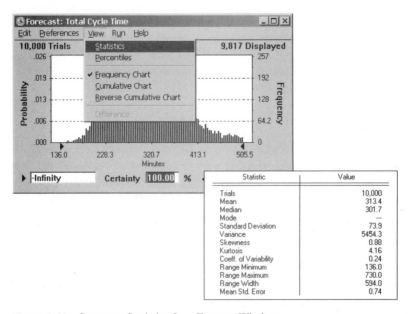

Figure A.10   Summary Statistics for a Forecast Window

1. Select "Run>Open Sensitivity Chart."
2. Click the "Chart Prefs..." button.
3. Make sure the "Total Cycle Time" line is selected and the "Contribution to Variance" option is selected (Fig. A.11).
4. Click "OK."

The analysis shows that the steps that contribute the most to the variation in total cycle time are the *returning delay* (34.3 percent), *rating delay* (22.8 percent), *distribution delay* (20.4 percent), and *pricing delay steps* (8.7 percent) (Fig. A.12). In general, the fact that they are all the delay steps is no surprise in the lean Six Sigma world. People are very consistent when they do a job, but the communication and handoffs are the most unpredictable and most prone to unexpectedly long times.

A.3.8   Extracting Data

Additional data analysis can be performed by extracting the data generated during the simulation (Fig. A.13).

1. Select "Run>Extract Data...."
2. Under "Type of Data," choose "Forecast Values."
3. Under "Forecasts," choose "All"
4. Click "OK."

A new tab will be created in the "Underwriting.xls" spreadsheet containing all the data from the simulation. Save the file in a convenient location for later analysis.

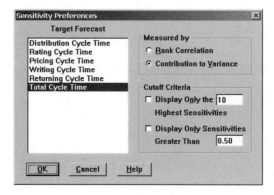

Figure A.11   Select Total Cycle Time and "Measured by Contribution to Variance"

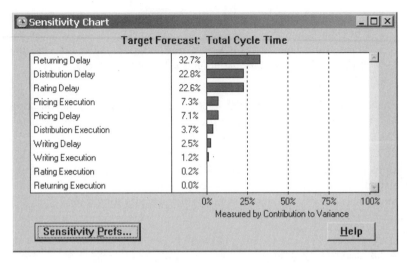

Figure A.12   Contribution to Variance for Total Cycle Time

A.3.9   Calculating Span from Extracted Data

The span metric is a measure of the width of a distribution that contains 90 percent of the observed data. The two extremes are defined as the fifth percentile (*P5*) and the 95th percentile (*P95*). In order to calculate this in Excel, find a few vacant cells in the "DATA" tab and enter the following: (Fig. A.14)

- =percentile(F:F,0.05)
- =percentile(F:F,0.95)
- =G3-G2

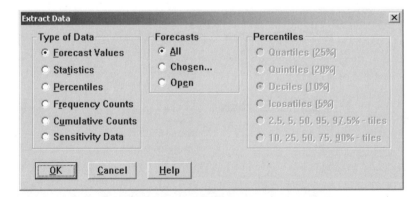

Figure A.13   Extracting Data After a Simulation Run

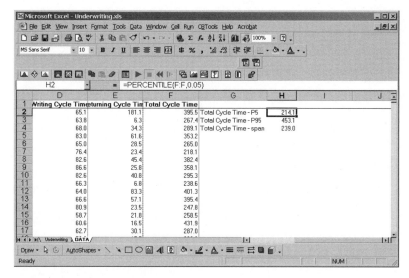

Figure A.14 Percentile and Span Calculations

Your span calculations will differ from these results owing to the random nature of the calculation, but should be similar. The variation in the process is such that some policies take over twenty times as long as others.

A.4   The Financial Impact of Process Risk—Medical Claim Payments

We are on the project team directed towards reducing the variation in time for processing medical claim payments. As part of the Measure phase we must assess the financial risk associated with the existing process. We have surveyed the customers and know that errors causing delays in payment are a common complaint and require a great deal of time on the part of the business to identify and correct. The belief from management is that the error rate is relatively low, on the order of 1 to 2 percent. Our job is to estimate the span of the problem in financial terms to baseline the process and get buy-in from management. The high-level process is shown in Fig. A.15.

We wish to determine:

1. The median and *P5-P95* span on interest expenses to the customers
2. The median and *P5-P95* span on the proportion of time spent in the business correcting errors

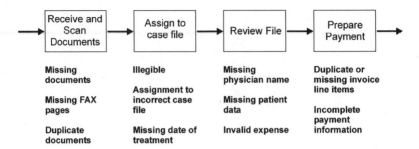

Figure A.15   High-Level Process of Medical Claim Payments

3. The sources of variation making an impact on the customer include:
   - Various interest rates they are assuming when making payments to their service providers before receiving their medical claim reimbursement
   - The time it takes to accomplish each process step
   - The size of the refund.

The sources of variation making an impact on the business are the ranges of:

- Time that the tasks take to execute
- Delay times to handle each input queue
- Time taken to correct any errors

During the data collection phase of the project, we used a check sheet to track the proportion of errors of different types and their associated execution and delay times. Historical data on size of claims was supplied by finance. The difficulty in summarizing the risk is that most of the claims were relatively small with a few very large claims.

The statistical summary for a sample of 1000 claims shows that the claim data is skewed to the right and follows a log-normal distribution, common for financial data. The distribution has a location parameter of 4.9 and a scale of 1.24 (Fig. A.16). A graphical summary shows that this data has a mean of about $300 with a standard deviation of about $515 (Fig. A.17).

A.4.1   Construct the Model

Construct the model using Crystal Ball as shown in Fig. A.18.

1. Start up Crystal Ball.
2. Enter the text for the column and row titles and set up the table in cells A1:L25.

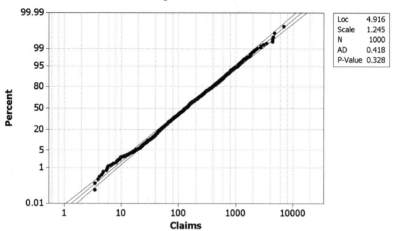

Figure A.16   Probability Distribution of Medical Claims

3. Enter the values for size of claims (B2) and interest rate (B3).
4. Enter the error rate information for the "Receive and scan documents" step in column B by entering the proportion of errors in the cells B7:B9 and using the formula "=1- sum (B7:B9)" in cell B6 to find the balance of nonerror related transactions for the "receiving and scan

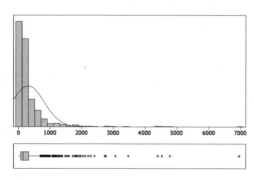

Figure A.17   Summary Statistics of Medical Claims

	A	B	C	D	E	F	G	H	I	J	K	L
1	**Medical Claims Processing**											
2	**Size of Claim($)**	$300.00										
3	**Annual Interest Rate**	10.0%										
4		Proportion	Delay Parameters (days)		Execution Parameters(days)		Delay Time	Execution Time	Time Value of Money	Error Correction	Normal Operations	Error/Total Time
5			Mean	Rate	Mean	Stdev						
6	Receive and scan documents	92.7%	0.2	5.00	0.1	0.03	0.2	0.1	$0.02		0.09	
7	Missing documents	1.2%	5	0.20	2	0.50	5	2	$0.01	0.02		
8	Missing FAX pages	3.2%	5	0.20	2	0.50	5	2	$0.02	0.06		
9	Duplicate documents	2.9%	2	0.50	1	0.25	2	1	$0.01	0.03		
10	Weighted Subtotal						0.46	0.21				
11	Assign to case file	95.4%	0.5	2.00	0.1	0.03	0.5	0.1	$0.05		0.10	
12	Illegible	0.5%	3	0.33	2	0.50	3	2	$0.00	0.01		
13	Incorrect case file	0.6%	14	0.07	5	1.25	14	5	$0.01	0.03		
14	Missing treatment date	3.5%	2	0.50	10	2.50	2	10	$0.03	0.35		
15	Weighted Subtotal						0.65	0.49				
16	Review file	62.5%	3	0.33	0.5	0.13	3	0.5	$0.18		0.31	
17	Missing physician name	0.2%	20	0.05	4	1.00	20	4	$0.00	0.01		
18	Missing patient data	2.1%	20	0.05	5	1.25	20	5	$0.04	0.11		
19	Invalid expense	35.2%	13	0.08	1	0.25	13	1	$0.41	0.35		
20	Weighted Subtotal						6.91	0.78				
21	Prepare payment	98.4%	5	0.20	0.1	0.03	5	0.1	$0.41		0.10	
22	Duplicate or missing invoice line items	0.5%	12	0.08	3	0.75	12	3	$0.01	0.02		
23	Incomplete payment information	1.1%	21	0.05	2	0.50	21	2	$0.02	0.02		
24	Weighted Subtotal						5.21	0.14				
25	Weighted Total						13.2	1.61	$1.22	1.01	0.60	0.63

Figure A.18   The Excel Model for Estimating the Financial Impact of Variation on the Customer and Time Correcting Errors on the Business

documents" step. Use the same logic to enter the remainder of the observed error rates for the "Assign to case file," "Review File," and "Prepare Payment" process steps.

5. Enter the mean delay times for each process step in column C. The delay rates for the error steps reflect the tendency for operators to correct errors at a lower priority than their regular work. The "Rate" column (D) is defined as the reciprocal of the entries in column C. For example, D6 is defined as "=1/C6." The rates are required to define the distributions for exponential delay in Crystal Ball.

6. Enter the mean and standard deviations of the execution times for each of the process steps in columns E and F, respectively.

7. Enter the values of the "Delay Time" and "Execution Time" in column G and H, respectively. Do not use cell references; enter the values themselves. These cells will be overwritten by Crystal Ball according to defined statistical distributions during the simulation and must contain single values, not formulae or cell references.

8. The TVM column is entered using the formula for continuously calculated interest.

$$\text{Interest} = \text{principal} \times (e^{\text{time} \times \text{rate}} - 1) \tag{A.1}$$

9. Enter the interest calculation for each item as a proportion of the number of times the claims proceed through the usual, correct process versus the longer, error-prone processes for each step. For example, the calculation for cell I6 uses the Excel formula, "=B2*B6*(EXP((G6+H6)/365*B3)−1)." Copy cell I6 and paste into cells I7:I9, I11:I14, I16:I19, and I21:I23.

10. Enter the weighted subtotal for "Delay Time" for the process steps by entering the formula "=$B6*G6+$B7*G7+$B8*G8+$B9*G9" in cell G10. Copy cell G10 and paste into cells G15 and G20. Use the formula, "=$B21*G21+$B22*G22+$B23*G23" in cell G24.

11. Enter the weighted subtotal for "Execution Time" for the process steps by entering the formula "=$B6*H6+$B7*H7+$B8*H8+$B9*H9" in cell H10. Copy cell H10 and paste into cells H15 and H20. Use the formula, "=$B21*H21+$B22*H22+$B23*H23" in cell H24.

12. Enter the "Error Correction" formulae. The calculation is merely the proportion of times a particular step is taken (column B) multiplied by the execution time for that step (column H). Cell J7 is "=B7*H7." Copy and paste cell J7 into the remaining error-prone steps in column J (rows 8–9, 12–14, 17–19, and 22–23).

13. Enter the "Normal Operation" formulae in column K. The calculation is similar to the one for the "Error Correction" column above. For example, cell K6 is "=B6*H6." Copy cell K6 and paste into cells K11, K16, and K21.

14. Enter the "Weighted Totals" in cells G25 for the "Delay Time," and H25 for the "Execution Time" columns by summing the subtotals in rows 10, 15, 20, and 24.

15. Enter the totals for columns I, J, and K in cells I25, J25, and K25.

16. The "Error/Total Time" in cell L25 is set to "=J25/ (J25+K25)." This number gives the proportion of time the staff spend correcting errors. It does not include any delay time in any of the steps.

17. Save the file as "Medical Claims Processing.xls."

The spreadsheet as it exists right now has only a single calculation of the quantities of interest. The customer complaints seem disproportionate with an expected value of $1.22 in interest. We are spending about 63 percent of our time correcting errors even though the error rate is quite low, in the range of 0.2 to about 2.5 percent. We need to investigate the effect of variation.

A.4.2   Define Assumptions

We will now modify the spreadsheet to incorporate the variation in the parameters by specifying their distributions. Enter the information from Fig. A.19 and Fig. A.20 by selecting cells one at a time and selecting "Cell>Define Assumption..." from the drop down menu.

A.4.3   Define Forecast Cells and Run Simulation

We have now defined all the parameters, relationships, and distributions for our model. It is now time to define the data we will be gathering. Select and define the following forecasts.

Select cell I25, select, and enter:

1. "Cell>Define Forecast..."
2. "Forecast Name: Time Value of Money"
3. "Units: Dollars"
4. "OK"

Cell	Distribution Type	Assumption Name	Parameters
B2	lognormal	Size of Claim($)	mean = $300, stdev = $580
B3	triangular	Interest Rate	Min = 3.5%, likeliest = 10.0%, max = 19.0%
G6	exponential	Receive Delay	rate '= D6'
G7	exponential	Missing Docs Delay	rate '= D7'
G8	exponential	Missing FAX Delay	rate '= D8'
G9	exponential	Duplicate Docs Delay	rate '= D9'
G11	exponential	Assign Delay	rate '= D11'
G12	exponential	Illegible Delay	rate '= D12'
G13	exponential	Incorrect Case Delay	rate '= D13'
G14	exponential	Treatment Date Delay	rate '= D14'
G16	exponential	Review Delay	rate '= D16'
G17	exponential	Physician Name Delay	rate '= D17'
G18	exponential	Missing Data Delay	rate '= D18'
G19	exponential	Invalid Expense Delay	rate '= D19'
G21	exponential	Prepare Delay	rate '= D21'
G22	exponential	Duplicate/Missing Delay	rate '= D22'
G23	exponential	Incomplete Info Delay	rate '= D23'

Figure A.19   Distribution Parameters for Claim Size, Interest Rate, and Delay Times

Cell	Distribution Type	Assumption Name	Parameters	
H6	Normal	Receive Exec	Mean '= E6'	Stdev '= F6'
H7	Normal	Missing Docs Exec	Mean '= E7'	Stdev '= F7'
H8	Normal	Missing FAX Exec	Mean '= E8'	Stdev '= F8'
H9	Normal	Duplicate Docs Exec	Mean '= E9'	Stdev '= F9'
H11	Normal	Assign Exec	Mean '= E11'	Stdev '= F11'
H12	Normal	Illegible Exec	Mean '= E12'	Stdev '= F12'
H13	Normal	Incorrect Case Exec	Mean '= E13'	Stdev '= F13'
H14	Normal	Treatment Date Exec	Mean '= E14'	Stdev '= F14'
H16	Normal	Review Exec	Mean '= E16'	Stdev '= F16'
H17	Normal	Physician Name Exec	Mean '= E17'	Stdev '= F17'
H18	Normal	Missing Data Exec	Mean '= E18'	Stdev '= F18'
H19	Normal	Invalid Expense Exec	Mean '= E19'	Stdev '= F19'
H21	Normal	Prepare Exec	Mean '= E21'	Stdev '= F21'
H22	Normal	Duplicate/Missing Exec	Mean '= E22'	Stdev '= F22'
H23	Normal	Incomplete Info Exec	Mean '= E23'	Stdev '= F23'

Figure A.20   Distribution Parameters for Execution Times

Select L25, select, and enter:

1. "Cell>Define Forecast..."
2. "Forecast Name: Error/Total Ratio"
3. "Units: Proportion"
4. "OK"

Select "Run>Run" to run 10,000 steps.

A.4.4   Evaluate the Results—Customer Impact

Your results will vary somewhat from the results shown here owing to the random nature of the simulation, but the conclusions will be similar. Figure A.21 shows that the financial impact on the customer has a very large range that is highly skewed to the right. While 50 percent of the customers accrue only about $0.53 in interest, the average is $1.34 with one customer having to pay as much as $115.

Select "Run>Open Sensitivity Chart" to see what is contributing to the large range. You may have to click the "Chart Prefs..." button and choose "Target Forecast: Time Value of Money" to see the desired chart. The output shows that the size of claim contributes about 85 percent of the variation (Fig. A.22).

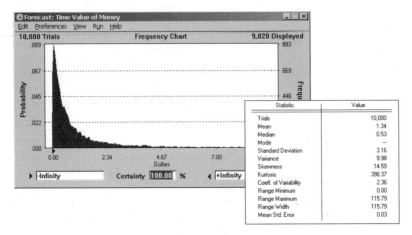

Figure A.21   The Range of Time Value of Money Values

If we wish to remove this factor, click the "Choose Assumption..." button in the sensitivity chart, click "B2: Size of Claim ($)" under the "Chosen Assumptions" column, then click the "<<Unchoose" button, and finally click "OK" (Fig. A.23).

Sensitivity Chart					
Target Forecast: Time Value of Money					
Size of Claim($)	84.5%				
Prepare payment delay	5.2%				
Annual Interest Rate	4.8%				
Invalid expense delay	4.2%				
Review delay	0.9%				
Assign delay	0.1%				
Receive delay	0.0%				
Missing patient data delay	0.0%				
Incomplete payment info delay	0.0%				
Incorrect case delay	0.0%				

100%   50%   0%   50%   100%
Measured by Contribution to Variance

Choose Assumptions...   Chart Prefs...   Help

Figure A.22   Contribution to Variance of Time Value of Money

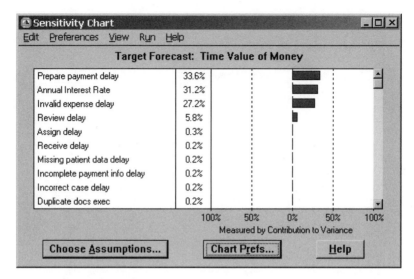

Figure A.23   Contribution to Variance of Time Value of Money Without Size of Claim ($)

The greatest opportunity for internal business process improvement would be to either reduce the delay for payment processing, or make it more consistent.

A.4.5   Evaluate the Results—Business Impact

Errors take a long time to correct. The impact of relatively low error rates can be much larger than anticipated. Select "Run>Forecast Windows...," select "Error/Total Time" and "Open Selected Forecasts" to find the forecast window (Fig. A.24).

Between 53 to 80 percent of the staff's time is spent on correcting errors in order to process the claims. The impact of errors on the business is even greater than that on the customers (Fig. A.25).

Select "Run>Open Sensitivity Chart," select "Chart Prefs..." and choose "Error/Total time" to see what is contributing to the large range (Fig. A.26).

Figure A.27 needs a bit of explanation. It shows that *variation* in the "review exec" process step has the largest contribution to the *variation* in error/total time. You could go back to the spreadsheet and define a "Forecast" cell for total delay time, total execution time, or total cycle time to determine the step contributing to the variation in each category.

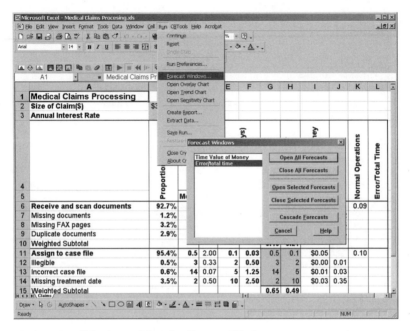

Figure A.24   Selecting and Opening Forecast Windows

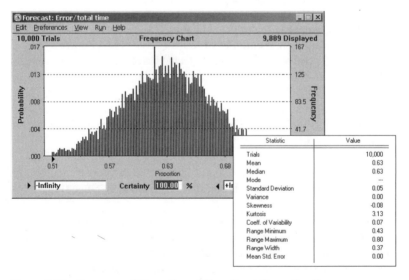

Figure A.25   Summary of Proportion of Time Spent Correcting Errors

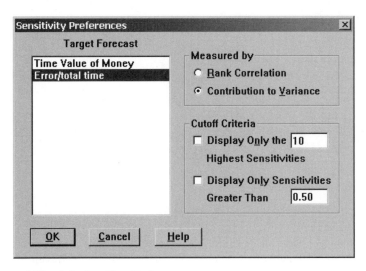

Figure A.26   Selecting Chart Preferences

A.4.6   Extract the Data and Calculate Span Metrics

Extract the data by selecting
- "R̲un>Extract D̲ata..."
- "Type of Data: All"

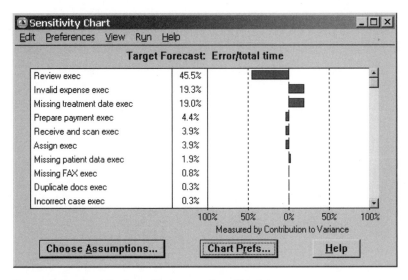

Figure A.27   Contribution to Variance to Error/Total time

- "Forecasts: All"
- "OK"

Determine the *P5-P95* span on TVM by evaluating the following in the newly created data sheet.

- "= percentile(A:A,0.05)." This should be about $0.05.
- "= percentile(A:A,0.95)." This should be about $4.98.
- The span on TVM indicates that 90 percent of the customers will accrue between $0.05 to $4.98 in interest.

Determine the *P5-P95* span on proportion of time spent correcting errors by evaluating the following:

- "= percentile(B:B,0.05)." This should be about 0.55.
- "= percentile(B:B, 0.95)." This should be about 0.70.
- The span on error time indicates that 90 percent of the time between 55 to 70 percent of the staffs' time will be spent correcting errors.

A.5 Summary

Crystal Ball can be used whenever there is a process where sequential steps are executed to give a total. The common assumptions of normally distributed processes are frequently violated for practical situations and the other tolerancing techniques such as "worst case" scenarios are overly pessimistic. The examples shown here show how Monte Carlo simulation can be used during the early Measure phase to support the business case by showing the *cost of quality* (COQ). This technique can also be applied during the Analyze phase to determine the vital *X*s of the process or the Improve phase to estimate the impact of different improvement strategies.

B

Process Simulation

B.1 Assessing and Designing a Transactional System

It is not easy to understand how typical transactional systems behave. Even well-designed ones are not linear assembly lines of productivity. We need a method to assess the behavior of systems that can incorporate:

- Multiple, nonconcurrent shifts
- Changing priorities for execution
- A wide range of execution times for subprocesses
- Transportation, splitting and consolidation of line items, shipping documents, approvals,and so on
- Complex rework loops
- Personnel with many different responsibilities

As lean Six Sigma becomes more and more engrained into the company culture, there will be a shift from a low volume, highly flexible custom job shop to a higher volume company that is much more standardized in product and service offerings. During the transition from the former to the latter, it is certainly not possible to physically try out all the different configurations of staffing, communication, and subprocess execution to determine the optimum process.

You could benchmark the process by investigating the application of some new technology by a competitor, but your business, staff, and customers are unique. Making a copy of another business' process is not going to give you the same kind of process performance they experience.

Instead of benchmarking, you could model the mathematical behavior of the subprocesses. When processes have simple, well-understood parameters for arrival rates, execution times and routings, then mathematical expressions can be used to derive quantities of interest. The average number in the queue, the average waiting time, and the standard deviation of the cycle time are all quantities that can be calculated. An example of simple system would be a few identical servers serving a single *first in first out* (FIFO) queue of entities with exponentially distributed interarrival times.

It is not possible to calculate the sorts of parameters we are interested in for the systems we are likely to encounter. In these cases, the best methodology is to construct a model with as many of the loops, range of values, and restrictions as possible and simulate the process using Monte Carlo techniques for thousands of entities. Data can be gathered for each of the entities and statistical analysis can be carried out after the simulation has run.

Models of processes have their limitations, but when used with care they can be used to identify bottlenecks and illustrate the unexpected reactions that can occur with a nonlinear, complex system. There are a variety of different software systems for handling models of different size and complexity. Our important requirements were

- Data export at the individual transaction level
- The ability to input and evaluate mathematical expressions to control and evaluate transaction processing during the simulation
- Flexibility to incorporate alternate routings and priorities
- Flexibility in the specification of distribution functions

We did not require centralized process management tools, upward compatibility with automated data gathering and control systems, simulation directed towards a specific industry, or 3D animation.

The applications we considered were

- Arena (www.arenasimulation.com)
- iGrafx (www.igrafx.com)
- Simul8 (www.simul8.com)
- ProcessModel5 (www.promodel.com)

B.2 Mapping and Modeling a Process

Creating a model of a process for the purposes of experimenting or optimizing the communication, task definitions, and resourcing is much more involved and detailed than the mapping exercise conducted for scoping the project, or for classifying steps as "value added" or "non value added." The workflow for defining a model of a process is:

1. Define the *entities*. This means the smallest unit of the transaction. If you are creating a model of order processing, then the smallest unit is

a line item. If you are modeling lab tests for a patient, then the smallest item is a particular test, even if the technician draws multiple samples at one time for multiple tests. There are times when a single transaction will create additional entities, such as when a loan application generates credit, employment, and reference checks.

2. Define the *activities*. Transactional processes are usually extremely complex where an individual activity may overlap with processes outside your project scope. If your process was scoped to include only domestic shipments, then parts of order processing will overlap with international shipments. While the preparation of customs documents is clearly outside your project scope, the preparation of shipping documents overlaps both types of shipments. This overlap has consequences with the availability of resources.

3. Define the *resources*. Most job functions in transactional systems are responsible for many, usually nonconsecutive, subprocesses. This creates constraints that both subprocesses cannot be executed by the same person at the same time, and that different subprocesses will be executed with different and changing priorities. Although it goes against the general principles of making the process flow, people will commonly batch activities that occur occasionally. Examples include making all cash calls once a week, grouping multiple orders to a single customer, performing reconciliations at the beginning of the month, posting orders at the end of the month, posting payments at the end of each week, and collecting samples from multiple patients in one run.

4. Enter detailed information on priorities, interarrival times, shifts, grouping, execution times, product, and service mix. It is common that a combination of grouping and reprioritization is done at the task level. Queues in transactional systems are commonly FIFO when a customer is directly in front of you, but usually otherwise for all other systems. There are a number of techniques discussed in Chap. 6 to characterize when people are maintaining multiple queues or when queue jumping is occurring.

5. Validate the existing model before making modifications. First, visually track the flow of a single entity of each transaction type throughout the system. This is to confirm both the flow logic and rework routing. Next, track the flow of many of the same type of entity to verify queue handling and rework priority. Finally, track the flow of many, different entities.

It is tempting to experiment with changing resources and priorities before the model has been validated. The best way to use the model is to conduct an analysis of "before" and "after" states. When a process is modeled, it may contain biases which make the results deceiving in an absolute sense.

When a change in a process is evaluated by comparing the difference between two configurations, then the biases tend to cancel each other out.

Rework loops will severely complicate a process map, but you must include them and realistically define the manner that they are incorporated into the upstream workflow.

B.3   ProcessModel5

B.3.1   The Interface

ProcessModel5 was chosen for modeling. The interface has a number of tool palettes for defining entities, activities, and resources. These tools are used to define the components of the process map by dragging and dropping them on the diagram workspace. One of the simple models we used for investigation during Chap. 7 was one where four workers were assigned to handle the customer requests. If the input queue for the workers exceeded three customers each, an overflow worker was assigned to the task and kept working as long as the queues were full. The four workers assigned to the normal process were loaded at 86 percent utilization while the single worker assigned to the overflow queue was loaded at 46 percent (Fig. B.1).

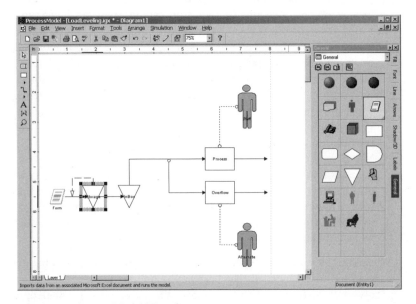

Figure B.1   ProcessModel5 Interface

B.3.2   Declaring Entity Attributes and Global Variables

The program automatically gathers some standard statistics on entities and process steps. We set the program up to gather a wealth of additional data on each individual transaction as it was passing though the process. This is done by creating named storage locations for extra parameters in the model. These parameters can then be manipulated at each process step. The differences between the two data types are:

- Entity attributes—properties relating to each individual entity passing through the system. Examples could include the individual priority of an entity, the individual start time, and end time in the process. In this manner, properties of the individual entities can be tracked as they move through the process. Some attributes are common and are defined by default (name, cost, VAtime, ID, cycle start).
- Global variables—properties relating to activities. Properties relating to variables can also be manipulated during the simulation. An example of a variable would be the number of entities in the input queue awaiting processing. The default variables are Qty_ processed, Avg_ VA_ Time, Avg_ Cycle_ Time, Avg_ Cost for each variable (Fig. B.2).

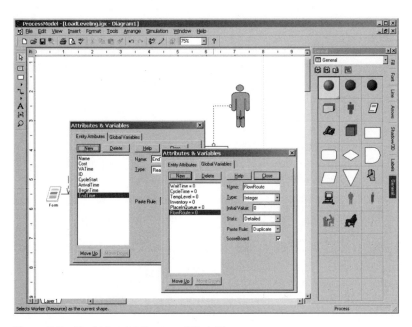

Figure B.2   Declaring Attributes and Variables

The chief difference between entity attributes and global variables is that the values of the latter are retained after a simulation and can be displayed as time series plots. Attribute data for individual entities are not retained for export.

Any data available for the time series plots can also be exported as a comma delimited file for detailed analysis in Excel or Minitab. An example would be to look at the variation in length of the input queues for subprocesses throughout the day. If entity data is required for later analysis, it must be transformed into a global variable during the simulation before it can be exported.

B.3.3 Action Logic

Action logic refers to statements and expressions that can be evaluated within the arrival, activity, and routing elements of the process map. Action logic can be used to write values of interest for entities (net waiting time) to global variables for retention and later analysis. The following is a summary of the logic used to track the individual waiting times. The extra global variables in the model are:

- WaitTime—the elapsed time an entity spends waiting for processing after arriving in the queue.
- CycleTime—the elapsed time an entity spends in the system. This includes both execution time and delay time.
- TempLevel—the number of entities that are waiting in the queue after leaving the inbox. This value is incremented as a new form arrives in the inbox and decremented as a form enters the process or overflow activities.
- Inventory—the number of entities either waiting in the input queue plus those being worked on.
- PlaceInQueue—the number of entities in the input queue as an entity enters the inbox.
- FlowRoute—a flag indicating which of the routine or overflow processes handled the entity.

As entities travel through the process, the values of the global variables are modified using the action logic at different points in the process.

Forms arrive at the storage element according to a Weibull distribution. As soon as the entity arrives, an arrival time stamp is assigned to the entity using the expression, "ArrivalTime = clock(sec)." In a similar manner the entities are time stamped when they enter either the process or overflow activities using, "BeginTime = clock(sec)" and finally when they leave the system using the expression, "EndTime = clock(sec)." Since the

ArrivalTime, BeginTime and EndTime attributes will disappear when the entity leaves the process, the global variables are used to calculate and capture the cycle time information for each entity. When each entity leaves the system, the two expressions, "WaitTime=BeginTime–ArrivalTime" and "CycleTime=EndTime–ArrivalTime" are evaluated.

A series of increments and decrements to the inventory variable are made to stamp each entity with a value that shows where it was in the queue for processing and how large the inventory level was when it arrived for processing.

B.3.4 Data Analysis

At the end of a simulation, the output module of ProcessModel5 summarizes some standard statistical parameters of the data. The output module can also be used to construct time series plots of the entire set of data for each global variable. We could also export the detailed transactional history of each entity as it passed through the process (Fig. B.3).

The output module can generate the time series plot directly, or the data can be exported to another program for plotting (Fig. B.4).

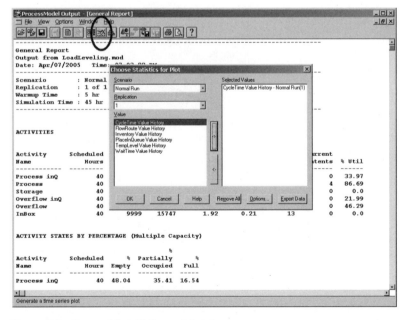

Figure B.3   ProcessModel5 Output Module

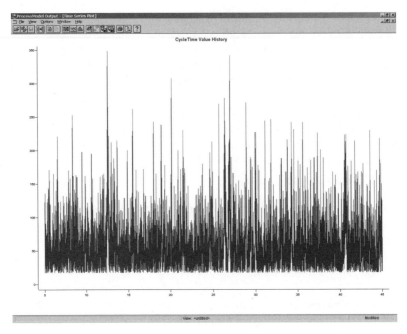

Figure B.4   Time Series Plot of Cycle Time

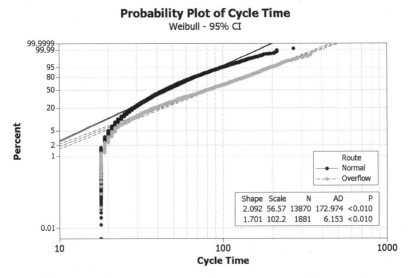

Figure B.5   Probability Plot of Cycle Time for a Transactional Process with Normal and Periodic Overflow Queues

The data analysis capability of ProcessModel5 is limited. The time series plot of Fig. B.4 shows the time behavior of cycle time without distinguishing whether the entities were processed by the regular or overflow queues. The data was exported to comma delimited files and read into Minitab. This allowed us to identify the probability distributions of the total cycle time broken out by queue type (Fig. B.5).

Whether you use ProcessModel5 or one of the other process simulation packages, follow the guidelines for constructing a rigorous model and make sure you have access to the raw output data for detailed analysis.

Statistical Analysis

C.1 Data Analysis

Lean Six Sigma is driven by data analysis. These requirements exceed the capabilities of spreadsheet software. *Master Black Belts* (MBBs), *Black Belts* (BBs), and lean Six Sigma project teams need to use the same software in order to share data, techniques, and results. Minitab (www.minitab.com) has emerged as the primary application in the Six Sigma community. The program assembles multiple sets of data, output from analyses, graphs, and printed reports in what are termed "project" files. Individual sets of data can be saved and shared between users by saving them as Minitab "worksheet" files.

C.1.1 The Interface

When Minitab is opened, the interface shows two windows. Figure C.1 shows the "Requests" Minitab project consisting of "before" and "after" worksheets.

1. The upper session window—the print out from analyses. When statistical analysis is performed, the output, for example, analysis of variance (ANOVA) table, is printed here. This text can be cut and pasted into presentations or added to the Minitab "Report" window. This example shows the output from the "<u>S</u>tat><u>T</u>ables><u>T</u>ally Individual Variables..." command.

2. The lower worksheet window—a tabular arrangement of data. It appears similar to a spreadsheet, but formulae can not be entered. Manipulations of data are performed using the "<u>C</u>alc><u>C</u>alculator..." function. The recognized data types are date/time, text, and numeric (default). This example shows the raw data for the project arranged in a "before" and "after" worksheet. The variables are:

   - C1-D (date format)–date/time stamp for a request
   - C2-T (text format)–the originating department
   - C3 (numeric format)–the number of items in the request

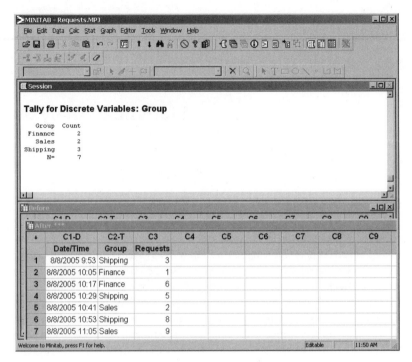

Figure C.1   Minitab Interface for the "Requests" Project

C.1.2   Project Manager

The files and variables associated with a Minitab project are managed using the *Project Manager*. Figure C.2 shows the status of the different entities in the project.

The example shows the project has two worksheets, "before" and "after." The open "session" folder in the left pane of the Project Manager window has been chosen to show more detail about the entities in the session window. The right pane of the window shows that the tally of individual values for "group" has been performed for both the "before" and "after" worksheets.

C.1.3   Data Input

Data can be entered directly into worksheets from the keyboard. It is quite likely that data will be imported from another application. The fastest method for transferring data into a Minitab worksheet is to cut and paste it

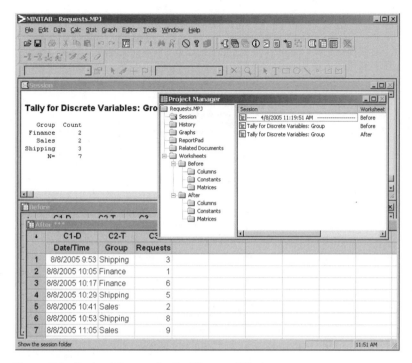

Figure C.2   Project Manager for the "Requests" Project

using the Windows clipboard. Comma delimited files generated from other applications can be imported directly into worksheets.

Remove all formatting from financial figures and check date formats before accepting the data. Minitab will sometimes split date/time stamp data into two columns.

C.1.4   Help

The Minitab "help" functionality is one of the best we have encountered. Most Minitab functions have a help button that explains the options available for the particular analysis being performed. Each function is explained using an example problem, complete with a data file (created when the program was installed) and the interpretation of the output.

The output in Fig. C.3 shows the help entry explaining application of ANOVA for testing whether there is a difference in the means of carpet durability with four different experimental treatments. The test file location and program

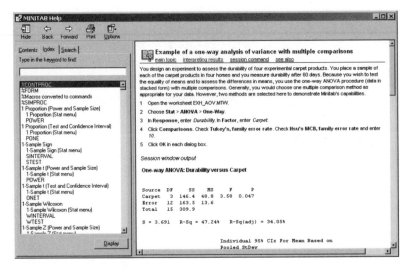

Figure C.3   Minitab Help Example

commands are listed. Most entries for examples in Minitab have an "inter-preting results" hyperlink detailing which results are important for assessing the results of the test.

In addition to the help file and examples, a hypertext statistics manual is available for most topics (Fig. C.4).

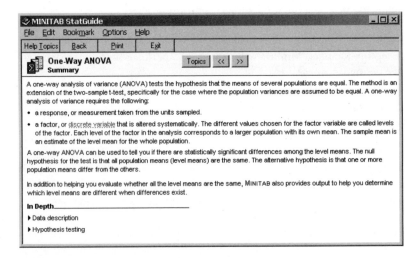

Figure C.4   Minitab StatGuide Statistics Manual

INDEX

ABOUT THE AUTHOR

Alastair Muir, president of Muir & Associates Consulting, is a leading Six Sigma consultant and author. His improvement projects in the manufacturing, service and consulting industries are driven by value creation and revenue growth.

He has been a Six Sigma consultant for GE businesses worldwide since 1997. Some of his other clients include Bombardier, AltaVista, EnCana, PricewaterhouseCoopers, and Diavik Diamond Mines. Dr. Muir, who lives in Calgary, Canada, has several degrees, including a Ph.D.

He can be reached at www.muir-and-associates.com.